Robot Building For Dummies®

Cheat Sheet

P9-DME-526

Expansion Connector Pinout

Expansion Connector Pin Number	Basic Stamp Port Pin	Signal Name
1,2		Ground
3,4		+5 volts
5	P0	Left whisker
6	P1	Right whisker
7	P2	Rear whisker
8	P3	Temp sensor
9	P4	Light sensor
10	P5	PIR sensor
11	P6	RC drive motor input
12	P7	Speech output
13	P8	Network to coprocessors
14	P9	Speaker
15	P10	Red LED
16	P11	Green LED
17	P12	Jumper J7
18	P13	Jumper J6
19	P14	Switch 1
20	P15	Switch 2
21		RB7 from coprocessor
22		Encoder pulse
23		Servo motor #1
24		Servo motor #2
25		Servo motor #3
26		Servo motor #4
27–40		Unused

Common Screw Drill Sizes

Screw	Tap Drill	Clearance Drill	Screw	Tap Drill	Clearance Drill
2-256	.070 (#50)	.089 (#43)	8-32	.136 (#29)	.170 (#18)
4-40	.089 (#43)	.116 (#32)	10-32	.159 (#21)	.196 (#9)
6-32	.106 (#36)	.144 (#27)			

For Dummies: Bestselling Book Series for Beginners

Robot Building For Dummies®

PBasic Programming Reference

For additional programming information, go to www.robotics.com/rbfd.

Reading whiskers

```
if in0=0 then wh1    'If left whisker on then jump to wh1
if in1=0 then wh2    'If right whisker on then jump to wh2
if in2=0 then wh3    'If rear whisker on then jump to wh3
```

Controlling LEDs

```
low 10     'Turn on the red LED
high 10    'Turn off the red LED
low 11     'Turn on the green LED
high 11    'Turn off the green LED
```

Controlling the speaker

```
freqout 9,200,1500  'High pitch beep
freqout 9,200,500   'Low pitch beep
'Increase 200 for longer duration
low 9               'Turn off speaker
```

Reading the buttons and switches

```
if in12=0 then j7    'If jumper J7 on then jump to j7
if in13=0 then j6    'If jumper J6 on then jump to j6
if in14=0 then s1    'If switch 1 on then jump to s1
if in15=0 then s2    'If switch 2 on then jump to s2
```

Drive motor and encoder

```
serout 8,396,["!M1160018"]    'Drive forward 12 inches
serout 8,396,["!M1060018"]    'Drive reverse 12 inches
serout 8,396,["!M1160048"]    'Drive forward 36 inches
serout 8,396,["!M1060048"]    'Drive reverse 36 inches
serout 8,396,["!M1100001"]    'Stop the drive motor
```

Use 'pause 100' after each motor command.

Steering motor control

```
serout net,baud,["!1R180"]    'Center steering
serout net,baud,["!1R1D0"]    'Steer left
serout net,baud,["!1R120"]    'Steer right
```

Use 'pause 300' after each steering command.

Head motor control

```
serout net,baud,["!1R280"]    'Center head
serout net,baud,["!1R2FF"]    'Head right
serout net,baud,["!1R201"]    'Head left
```

Use 'pause 300' after each head command.

For Dummies: Bestselling Book Series for Beginners

Robot Building FOR DUMMIES®

by Roger Arrick

WILEY

Wiley Publishing, Inc.

Robot Building For Dummies®

Published by
Wiley Publishing, Inc.
111 River Street
Hoboken, NJ 07030

www.wiley.com

Copyright © 2003 by Wiley Publishing, Inc., Indianapolis, Indiana

Published by Wiley Publishing, Inc., Indianapolis, Indiana

Published simultaneously in Canada

For general information on our other products and services or to obtain technical support, please contact our Customer Care Department within the U.S. at 800-762-2974, outside the U.S. at 317-572-3993, or fax 317-572-4002.

Wiley also publishes its books in a variety of electronic formats. Some content that appears in print may not be available in electronic books.

Library of Congress Control Number: 2003105857

ISBN: 0-7645-4069-6

Manufactured in the United States of America

10 9 8 7 6 5 4 3 2 1

1O/RW/QZ/QT/IN

WILEY   is a trademark of Wiley Publishing, Inc.

About the Author

Roger Arrick is founder of Arrick Robotics (www.robotics.com), a small manufacturing firm based in Texas that designs and builds factory automation products along with mobile robots for use by hobbyists and educators. Roger is also past-president of the Dallas Personal Robotics Group (www.dprg.org), one of the largest and oldest robot special interest groups in the nation, and founder of the East Texas Robot Builders Group (www.etrb.org).

Author's Acknowledgments

Like the rest of the text in this book, I struggled over the content of this section. Albeit small, it's outside the grasp of certain editorial axing techniques, and I find that as liberating as running naked through a large wheat field with scissors in hand. The smile you might see on my face is likely from strange sensations, not joy.

The first person I'd like to offer thanks to is actually a fictional person named Homer Price who, through the writing of Robert McCloskey, showed me early in life that a little boy with enough energy and imagination could make more donuts than anyone could ever eat. Yes, gears were involved, and that short story helped set mine in motion.

I'd also like to thank the editors, especially Melody, who endured the whining of a would-be writer in over his head in more ways than one. Susan and Nancy also deserve a special thank you and a gift certificate to Vegas for six months. I can probably arrange one of those items.

Thanks are also due to my family, friends, and employees, who endured what seemed like decades of missing time, lost opportunities, and the senseless tossing of perfectly good computer monitors — all while providing a much needed sounding board for issues that professional writers have learned to endure with a grin.

Special thanks to Jim Brown for his writing, wit, and endurance. The secret robot builder's handshake is in order.

And thanks to Midi for his unwavering companionship, often expressed by spasmodic jumping, 40-mile-per-hour dashes to nowhere, and freshly dug holes large enough to shelter a riding lawnmower. He seemed unfazed by this whole process, and that is what I appreciate him for.

And lastly, yet certainly firstly, to our Creator, who makes this all possible and provides all these opportunities for reasons I'll never fully understand.

Publisher's Acknowledgments

We're proud of this book; please send us your comments through our online registration form located at www.dummies.com/register/.

Some of the people who helped bring this book to market include the following:

Acquisitions, Editorial, and Media Development

Project Editor: Susan Pink

Acquisitions Editor: Melody Layne

Technical Editor: Jesse Torres

Editorial Manager: Carol Sheehan

Permissions Editor: Laura Moss

Media Development Supervisor: Richard Graves

Editorial Assistant: Amanda Foxworth

Cartoons: Rich Tennant (www.the5thwave.com)

Production

Project Coordinator: Erin Smith

Layout and Graphics: Seth Conley, LeAndra Hosier, Stephanie D. Jumper, Mary Gillot Virgin, Shae Lynn Wilson

Proofreaders: John Greenough, Andy Hollandbeck, Carl William Pierce, Brian H. Walls, TECHBOOKS Production Services

Indexer: TECHBOOKS Production Services

Special Help: Laura Bowman

Publishing and Editorial for Technology Dummies

 Richard Swadley, Vice President and Executive Group Publisher

 Andy Cummings, Vice President and Publisher

 Mary C. Corder, Editorial Director

Publishing for Consumer Dummies

 Diane Graves Steele, Vice President and Publisher

 Joyce Pepple, Acquisitions Director

Composition Services

 Gerry Fahey, Vice President of Production Services

 Debbie Stailey, Director of Composition Services

Contents at a Glance

Table of Contents

Introduction

M aybe your first exposure to robots was in an unmemorable 1950's B movie titled something like *Amazon Women on Mars*. Maybe you played with remote-controlled cars as a kid and always loved having little mechanical things zipping around your house. Whatever the origin of your impulse, if you picked up this book you probably have the robot bug. The next logical step is to start building robots of your own.

Consider this your robot cookbook.

About This Book

This book is for the tinkerer, the dreamer, and the robot builder in us all. I know creative types like you don't want to wade through stacks of books or jargon to get that first robot built, so I cut to the chase: You get just what you need in plain English to get your own robot production line up and running — and, most importantly, you'll have fun doing it.

In this book you get an overview of robotics and the tools, technology, and skills you need to become a robot builder. You discover the various approaches to robot building, such as building from scratch or starting with a kit. You also explore the various mechanical parts of a robot and how they fit together, as well as programming basics you need to enter and download commands into your robot. Finally, you find out how to add functionality such as motion detection and light sensors to your robot.

Conventions Used in This Book

Book publishers have learned a thing or two about getting information across to readers in a logical fashion. That's why you'll find a variety of text formatting in this book:

- Web addresses look like this:

    www.robotics.com

- When I use an unusual term for the first time, I *italicize* it.

✔ Programming code looks like this:

```
'-----------------------------------------------
'testrw.bs2
'
'This program tests the rear whisker.
'When the whisker is activated, the red LED will
'turn on.

testrw: if in2 = 0 then testr    'Check rear whisker
        high 10                  'LED off
        goto testrw              'Loop forever

testr:  low 10                   'LED on
        goto testrw              'Loop forever
```

What You're Not to Read

Some information in this book you'll probably have to read in order. For example, you should read about setting up your robot building workshop before you read about building the ARobot. You should read about building a robot before you read about programming the robot.

Then there's information you can skip entirely, depending on what you already know. If you know all about computer programming, for example, you can skip the chapter on programming. If you know all about electronics, you can skip the sections on electronic basics.

After you build your robot, you can dip into the chapters in Part IV, which provide a variety of additional functionality to your robot, in any order you want. You can do a project from Part IV today, put the book aside, and come back a few months down the road and do another project. If you have no intention of ever adding a particular function, such as speech, don't read that chapter. (But you'll be missing out on something: Conversation with a robot can be very stimulating!)

Foolish Assumptions

If you're a bit daunted by the idea of building a robot, you'll be glad to hear that I haven't assumed that you have a lot of knowledge about things such as electronics, high-tech tools, or computer programming. I assume that you can work with some basic tools such as pliers, a screwdriver, and a cutting knife without running to the hospital emergency room every few days.

I also assume that you have a computer and know your way around, even though you may never have written a computer program in your life. In many cases I refer you to online information or sources, so I hope you have access to the Internet. (Remember, you can always go to your local library for Internet access.)

You need some kind of workspace, even if it's only your kitchen table or a space on your living room floor, and you should be able to keep small parts and sharp tools safe from younger members of your household and pets.

Other than that, all you need is enthusiasm, a bit of patience, and a little money to buy robot-building supplies.

Note: This book uses the ARobot kit to walk you through building and programming a robot. Much of the information and programming advice is applicable to other, similar kits. You will have to purchase such a kit to work through the chapters on robot building and programming. These types of kits and a controller will run you between $200 and $300. A smaller, more afford-able, nonprogrammable robot called Soccer Jr. is also introduced in this book and costs less than $50.

How This Book Is Organized

Building anything suggests a logical progression, so in this book I've tried to help you by organizing chapters and parts of the book in a logical way. Essentially, you go from basic concepts, to a discussion of the tools, hardware, and software you'll need to create and animate your robot, to hands-on projects for adding functionality to your robot.

Part 1: Getting Started with Robotics

In Chapter 1, you read about the state of robotics today — where it started and how far we've come. You discover how robots are being used in the real world and what types of activities robot building involves.

In Chapter 2, I discuss various ways to build robots, from making one from scratch to using a predesigned kit. (These are the equivalent of gathering together flour and sugar and whatnot to bake a cake versus making the acquaintance of a lady named Betty Crocker and just adding water.)

In Chapter 3, I introduce an affordable, nonprogrammable robot. A small project like this will help you hone your building skills and acquaint you with robotic terms and ideas.

Part II: Programmable Robot Prep

Before you can run, you have to walk. Part II provides a stroll through the preparations involved in robot building.

Chapter 4 guides you through setting up a robotics workshop, including finding the right space, and assembling tools, storage, shelving, and even a library of information and parts catalogs.

Chapter 5 goes over the various hardware items you use in building robots, from nuts and bolts to motors and software.

Chapter 6 is a handy, dandy programming primer, giving you the basics you need to know to enter simple programs in an editor and download them to your robot.

Part III: Building a Programmable Robot

In Part III, you get down to assembling your first programmable robot. This can be an exciting time, but to get things right you need to proceed step by step.

In Chapter 7, you take the time to prepare yourself and your robot kit materials, trimming rough edges and sanding and painting the robot body.

Chapter 8 is where you assemble the chassis, wheels, and whisker sensors that make up ARobot.

Chapter 9 introduces you to the editor software you'll use on your computer to enter programs for your robot.

Chapter 10 is where you'll finally make the connection between computer and robot, stringing cables and downloading programs that will get your new robot buddy going.

Part IV: Augmenting Your Programmable Robot

Your favorite part of the book might be Part IV because this is where your new robot gets up and does stuff.

Chapter 11 sets the stage for the other chapters in this part by walking you through the process of adding an expansion board to your robot (an expansion board is where you attach all the sensors, cameras, and other stuff used to help your robot interact with its environment).

Chapter 12 is where your robot learns about light and dark as you mount a light sensor to it and download a program to measure light.

Chapter 13 takes you through the process of installing a temperature sensor on your robot. With this sensor in place, your robot can take temperature measurements in your basement, a live volcano, or wherever! (But I'd think twice about the live volcano part. . . .)

Chapter 14 opens your robot to the world of motion. By installing a motion detector and downloading various programs to your robot, it can alert you when intruders are present or just monitor the activities of your pet dog.

Chapter 15 explores speech technology and is where you can enable your robot to speak, whether it has something to say or not (like many humans I know).

Chapter 16 is where you add vision to your robot, allowing it to take images of its surroundings and send them to your monitor.

Chapter 17 covers remote-control technology that you can add to your robot to control it from afar.

Part V: The Part of Tens

The chapters in the Part of Tens give you quick hits of information on a variety of topics. In this case, Chapter 18 gives you the lowdown on some excellent robot parts suppliers (ten, of course). Chapter 19 covers ten important safety pointers for robot builders.

Icons Used in This Book

Icons are little graphic things sprinkled liberally around this book to grab your attention. Each one alerts you to a slightly different type of information.

The tip icon tells you that some extra tidbit of advice or wisdom is coming your way.

A remember icon is a reminder, suggesting something you may have discovered in another chapter or a point that's well worth memorizing — it's that important.

Hard as I try, occasionally I go into technical-speak. If you're a techie, read it. If you're not, you probably will do fine if you don't.

Warnings are not optional reading material: These items alert you to potentially dangerous situations. When you're working with electricity and sharp tools, as you do in robot building, you'd be wise to read each and every warning.

Where to Go from Here

It's time to enter the brave new world of robots. I think you'll be amused, amazed, and intrigued by what you can do with robotics with just a few tools, some parts, and a lot of imagination.

Enjoy the journey!

Part I
Getting Started with Robotics

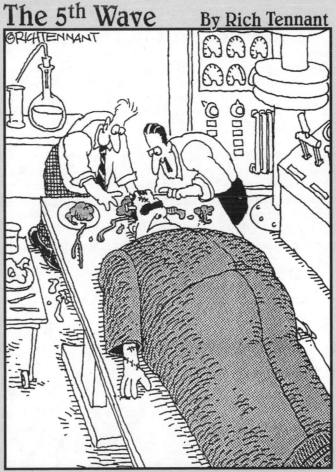

The 5th Wave   By Rich Tennant

"Wait! Wait! Wait! You've got a lung and two eyeballs in there! This is too hard. We should've started with a simple robot instead."

In this part . . .

The longest journey begins with a single step, and your robot-building journey is about to begin. To become a bona fide robot builder, your first step is to understand a bit about the world of robotics and what options are available to you.

In this part you get a look at the state of robotics today — how robots are being designed, built, and used. I explain the options for building robots from scratch or from a variety of kits. Finally, you start your robot-building adventure by assembling a simple nonprogrammable robot. You can't download computer programs to it. Instead, it performs a single, preprogrammed function.

Time to begin. Ready, set . . . go!

Chapter 1

So, You Want to Build a Robot?

In This Chapter

▶ Introducing the robotics revolution

▶ Finding out what a robot can do for you

*E*ver since the early days of science fiction, people have dreamed of little metal men running around doing our chores, alerting us to danger, performing the dirty jobs (such as going into deserted alien mine shafts on Mars to measure deadly radioactive rocks), and generally embodying technology in a quasi-human form.

How far have we come from these early fantasies? Quite a ways, but much progress is still to be made — maybe more than you think. Because our idea of robots comes mainly from the big screen, our view is skewed. Instead of robots doing our dishes and playing basketball with the kids, today's robots are mainly used to automate repetitive and boring manufacturing tasks and in dangerous jobs, handling bombs or radioactive material.

The lack of movie-style robotic technology may make you think that the fun stuff is so far away that it's not worth pursuing. Nothing could be further from the truth. In fact, now is the time to jump on the bandwagon and get involved — you'll be able to tell your grandkids that you were one of the pioneers.

Robotic technology is in its infancy, just like personal computers were in the '70s. Back then, computers were built by a few special people known as geeks, they cost a lot of money, and they could barely balance a checkbook. Robotics will be one of the next great technological revolutions, and you're getting in on the ground floor!

In this chapter, you find out about the state of robotics today, elements you'll encounter when working with robots, and some cool uses for robots. So come with me on a tour of robot world, and prepare to be amazed.

The Robotics Revolution

Welcome to the robotics revolution. You're reading this book, so you're already intrigued by robotics. But to really become a card-carrying revolutionary (robot builders unite!), you should understand a bit about where all this interest in robots comes from and where it's going.

Where have we been?

Computers are used to program robots, so it's not surprising that the growing interest in robotics is tied to the rapid advancement of computer technology.

It used to be that only the U.S. government, large corporations, or major colleges could afford to own a computer. Over time, computers got cheaper and smaller until finally even hobbyists could afford to own one. With all these cheap computers floating around, it wasn't long until robotic kits began to appear. Kits such as Heathkit's Hero robot kit brought robotics into the age where everyone could own and build their own autonomous personal robots, complete with real computer brains. And they could download programs to control their robots from their handy desktop PC.

That's when robot clubs such as the Dallas Personal Robotics Group and the Seattle Robotics Society started popping up all across the country. Robot fever had begun. People with their new robot sidekicks wanted to talk to other people who had the same interests, and the availability of Internet newsgroups, discussion boards, and chats fueled the robot fire. Robotics groups grew in number and in size. Many robotics groups with huge member lists became modest organizations equipped with newsletters, Web sites, and e-mail lists. Nowadays, you can pretty much find a robotics group in every nook and cranny of the country. Bottom line: A lot of people want to build robots these days.

Are we there yet?

A robot is simply any mechanical unit that detects its surroundings, is capable of decision making based on input about those surroundings, and performs some operation based on those decisions. In many ways, we have arrived in a world where robots are a reality. We don't have a humanoid-looking robot helper serving cocktails in every house yet, but robotic devices have been installed in many homes across the United States. You may have one lurking in your own house and not even know it.

For example, you might not think of motion detectors and electronic thermostats as household robots, but both devices sense their surroundings, make electronic decisions, and perform some action based on what they sense. Likewise, dishwashers and clothes washers look more like big white boxes than like robots, but their combination of electronic and mechanical operations make them at least distant cousins to robots. VCRs and DVD players have sensors that read media as well as computers that make decisions and adjustments many times a second to control motors and video output. Personal computers, phone answering machines, and clock radios could be considered robotic. Take a look around your house, and you may find other electronic devices that perform helpful robotic chores without your intervention.

But household appliances aren't really where it's at for those of us who live, eat, and breathe robots. Most people want to see a robot walk, talk, flash alarm lights, or roll around on its own. If you're one of those people, you'll be glad to know that several "real" robots are available today. For example:

- ✔ Large robotic couriers prowl the corridors of many of our hospitals, ferrying blood samples to medical labs and intelligently avoiding objects in their paths.

- ✔ A robot called Wisor lurks beneath the streets of New York City, repairing old steam pipes in places where people would just as soon not go.

- ✔ A remote-controlled robot named Predator was used by the U.S. military for surveillance duties in Afghanistan.

- ✔ Federal emergency teams have robots that can move between open spaces in rubble to help find trapped earthquake victims.

- ✔ NASA uses the MER rovers *Spirit* and *Opportunity* to explore Mars.

Not all robots are developed by large corporations or the government. Many robotic devices begin as the daydream of a hobbyist, who then becomes an entrepreneur and goes on to create a company to make and sell robots. What was once a vacuum-cleaning robotics game is now a viable household appliance. What was once a robotic builder's dream and a robotic group's ongoing hypothetical discussion is now a robotic lawnmower. What was once a geek's plaything is now a doctor's telepresence in the form of a geriatric monitoring device.

Many companies offer marketable robot products including do-it-yourself kits for the robot enthusiast. Robots are beginning to make it into the mainstream, albeit slowly.

Where are we going?

In twenty years, people will look back at the movies of today and laugh at what we thought robots would be like in the future. We've all heard the promises of a robot in every house and flying cars zooming around with a robot at the wheel, a la George Jetson. We're all waiting for the day when a robot will do all the household chores. But aren't these just fantasies with no more foundation than rumors of Elvis's latest sighting?

Actually, all these possibilities could come true in the not-too-distant future. With so many people getting involved in robotics and with robot parts and electronics getting cheaper day by day, you're likely to see robot advancements in just about every area of life.

In the future, you could see humanoid robots doing all the work. But, who says robots have to look humanoid? Your car could have a computer brain installed and become the robot that drives — no, flies — you to work. Perhaps clothes could be a robotic exoskeleton that helps you lift heavy objects or helps the physically challenged to walk. The U.S. military, for example, is experimenting with specially enhanced robotic combat fatigues (as reported by robotic news sites such as `www.robots.net`).

In the future, whatever their form, robots will be around at your beck and call to assist you and serve you.

Robot Uses

The word *robot* first appeared in the English language in 1923, and comes from the Czech word, *robata,* meaning drudgery or servitude. Early robot enthusiasts clearly saw robots as mechanical servants meant to give us a life free from the more mundane tasks.

So if robots aren't yet flying us to work or washing the dog, what are they doing? Today, robots are being used more and more for jobs that are too tedious or too dangerous for people to do, such as fitting parts on assembly lines or sensing landmines in war zones. Robots are also going to places that are too distant and dangerous for humans, such as the crushing depths at the bottom of the ocean, highly radioactive areas, and the hostile environments of Mars. In this section I take you on a tour of some intriguing robot activity.

Security

If Mr. Worf had had a robot to handle security, the Starship Enterprise would have been a different place. No Klingon angst. Just good, functional, robot security patrols.

Today, several types of robots can perform basic security functions. Robots roam the halls of museums and detect movement, humidity level, and fire.

Robots handle entry management for secure buildings. Airports have robotic devices that scan luggage for bombs, and robotic cameras that can do retinal scans and perform face recognition analyses. Police departments use remotely-operated robots for disarming bombs and negotiating with potentially dangerous criminals (put down your gun . . . beep).

Because security involves both the tedious (endlessly walking the halls of buildings at night) and the dangerous (disarming bombs and drawing gunfire), robots and robotic devices are a vital part of today's security force and will be even more so in the future.

Surveillance and exploration

Robots today go where no man has gone before, from the top of a volcano to the wreck of the Titanic in the depths of the ocean. Why are robots showing up in these odd spots? They go there to perform surveillance operations no one in their right mind would try to do.

NASA and other space agencies have found that it's cheaper and smarter to use robots to explore our solar system. To keep an eye on our own planet from space, robotic spy satellites view and remotely monitor the earth's surface from hundreds of miles above the earth. *Spy bugs,* as these satellites are called, have tiny legs (making them truly look like bugs) and use tiny color cameras to view their surroundings.

Contests

What's the point of building a little robotic buddy if you can't make a sport of it? More than thirty years ago, the first robot contest occurred in a hallway at MIT. The subject of a mechanical engineering class, this experiment was the precursor to today's robot contest boom.

Most robotics groups hold contests that range from simple robotic navigational contests to robotic vacuum-cleaning contests and robotic fire-fighting contests. Robot contests are exciting and draw members into robotics clubs like flies to honey. Once hooked, members drag in friends and relatives and the sport just gets better as the number of participants grows.

National contests sponsored by those ranging from wealthy technology entrepreneurs to the U.S. government draw tens of thousands of enthusiasts, sometimes for large cash prizes. For example, in 2004, the military is sponsoring a robot race from Los Angeles to Las Vegas with a grand prize of one million dollars — all to inspire new ideas about how robot-controlled vehicles might navigate rough terrain in military operations.

Robotic contests serve useful purposes: They help to further the acceptance and growth of robotics, stimulate research into new robot technology, and educate students about technology in a fun and creative way.

Your own grocery store may monitor its parking lot with remote cameras that can pan in response to movement. Some home security systems can monitor the front doorstep or the back driveway and send alerts if a presence is detected.

In short, robots are great tools for observation, and robotic surveillance has become commonplace. As this robotic application becomes more prevalent, there could be some bumps along the way — namely, issues with human privacy.

Human helper

You probably won't be surprised to hear that several robots are already ready and willing to help out with those pesky household chores. Currently, robots can vacuum your floors, mow your lawn, and more. Most of these robots are not humanoid at all, but rather resemble low-to-the-ground golf carts.

Because your average person likes a human face on a machine, much research is going on in the area of humanoid helper-type robots, from figuring out how to make a robot such as Honda's Asimo walk, to giving robots human-like facial expressions. Mitsubishi's Wakamaru, a home-caregiver robot now in development, is rumored to be capable of everything from sending e-mail to a family member in an emergency to giving its human charge a great big hug.

As robots and humans interact, various social, economic, and safety issues arise. Is the Sony Aibo dog a substitute for a real pet? Do robots make us lazy? How can robots help — or hinder — our humanity and human relationships? Robots haven't advanced so far that we need to stay up nights worrying about their effects on us, but it's a good idea to keep in mind Isaac Asimov's Revised Laws of Robotics:

- ✔ **Zeroth Law:** A robot may not injure humanity or, through inaction, allow humanity to come to harm.

- ✔ **First Law:** A robot may not injure a human being or, through inaction, allow a human being to come to harm, unless this would violate the Zeroth Law of Robotics.

- ✔ **Second Law:** A robot must obey orders given it by human beings, except where such orders would conflict with the Zeroth or First Law.

- ✔ **Third Law:** A robot must protect its own existence as long as such protection does not conflict with the Zeroth, First, or Second Law.

Building robots introduces us to this brave new world.

Chapter 2

Plotting a Path

*E*very learning experience is a kind of journey. Now you wouldn't set out to travel from New York to Chicago without a map, would you? You shouldn't set out on your robot knowledge quest clueless, either. So first, get a plan.

Some of you may want to jump right in and build a robot from scratch. That's like inexperienced travelers choosing to travel to a war zone for their vacation, so I don't recommend it. What I do recommend is that you start out simple with a few basic robotic kits. Then as you grow in skill and knowledge, you can advance to more complex robots and more complex construction techniques.

In this chapter, I show you some possible robot construction paths, help you choose your itinerary, and map out your journey.

Starting with a Kit

By now, you're probably one of us — you know, the elite group of people who have been bitten by the robot bug. Although this particular bite doesn't itch, it does give you a strong desire to be around robots, play with robots, and get one-on-one with robots. This means you're ready to dig in and build your own, which is great. But exactly how to start is not so obvious.

Why not just build your own robot buddy out of scrap parts lying around the garage? After all, you have a vague idea about what's inside a robot, and most of those things are rolling around in the bottom of your junk drawer or in a plastic bucket in the garage. Any parts you don't have you could easily get from a hardware store or a mail-order supply house.

Trust me, building a robot from scratch is not the best place to start. To transform common household items into useful components that actually fit together usually requires a drill press, a milling machine, and a welder. And you'd still have to buy plenty of components that aren't likely to be lying around the house. The process of building a robot from scratch requires a good design, a healthy dose of knowledge and skills, more time than most of us are willing to commit, and, frankly, a bucket full of money.

A better route is to use a robot kit. With a kit, some other poor soul gets to do the measuring, drilling, milling, and design. You get to have the fun of putting together a working robot. It might not be capable of exploring the surface of Mars, but it would be a good springboard for your next project. Putting together a kit also familiarizes you with robot terminology and components and gives you some ideas for making improvements when you build your next robot. After all, building robots is like the universe: constantly expanding.

Robotic kit companies try to reduce the complexity of building robots so that you don't need a Masters in Electronic Engineering and a crew of Ph.D. buddies on hand. You'll find that with many robotic kits, you can tackle building a robot all by yourself.

Even if you can build a kit by yourself, you may want to work with a friend or get input from folks in a robot club in your area if need a little help: Two heads are always better than one.

Selecting a Robot Kit

Now that I've made a compelling argument for crawling before sprinting, you have to select a robot kit. The first thing to consider when choosing your robot kit is whether you want to work with a nonprogrammable or programmable kit.

Of the many robot kits available to the hobbyist, most are designed to perform only one hardwired function, such as following a line or moving in response to a sound. This limited capability is mainly because inexpensive robot kits usually don't have a computer controller that can be programmed, hence the term *nonprogrammable robot*.

Wait, these ROVs really are robots?

Just when you think you know what a robot is, someone throws a wrench into the works. There's a difference between a robot that requires human interaction (an ROV) and one that 's autonomous. But then there are robots in-between that are a little of both. Some ROVs operate autonomously to some degree, and some robots require a little human intervention.

An ROV such as an RC helicopter, for example, maintains its balance autonomously while the operator tells it where to go. And the Mars Rover robot frequently had to wait for instructions from mission headquarters before continuing with its autonomous mission. It's hard to say what is and isn't a robot, so you're better off not even trying.

You can purchase nonprogrammable robot kits online, from catalogs, and even at toy stores and technology specialty shops. Prices range from $25 to $100. They're usually made of plastic and run on normal household batteries. You begin your foray into robot building by assembling a nonprogrammable robot in Chapter 3.

Robot kits are great fun and helpful for the beginning robot builder, but don't expect them to do much more than their predetermined task. It's rare for a low-end robot kit to allow any expansion, but it's common to see their components used to build more advanced, custom robots as the robot builder gains experience.

In Part II, you work with a programmable robot called ARobot. The benefit of a *programmable robot* is that you can — duh — program it. With programming, there's no end to the stuff you can get your robot to do: make sounds, navigate around objects, flash lights, sense motion, record temperatures, and on and on.

It's rare for a low-end robot kit to allow any expansion, but it is common to see their components scavenged to build more advanced, custom robots. You can scavenge a few parts yourself when you get more experienced.

Nonprogrammable kits

As mentioned, the most basic starter robotic kits are typically nonprogrammable robots. One good example of an easy-to-use nonprogrammable kit is the Soccer Jr. robot from OWI, Inc. (www.robotkitsdirect.com), which is shown in Figure 2-1. In fact, I like it so much that I chose it for the first robot-building project (see Chapter 3).

Figure 2-1:
The nonprogrammable Soccer Jr. robot.

The little plastic Soccer Jr. robot not only performs a task but also offers some human interaction and control. The kit comes with a wired controller that allows it to move in any direction and capture and shoot small soccer balls (well, they're actually ping-pong balls). You can even enter it in certain robot competitions.

The Hyper Line Tracker, shown in Figure 2-2, is an intermediate-level nonprogrammable robotic kit from OWI Kit. Unlike Soccer Jr., The Hyper Line Tracker requires no constant human interaction. Instead, it performs one preprogrammed task: following a line. You get to draw the line, which might be more fun than you think. Line following is not a useless task; many industrial robots that manage warehouses use a similar concept to navigate.

When choosing a nonprogrammable robot kit, start with a simple robot that you're sure you can tackle and then progress to more advanced robots as your confidence and knowledge increases.

You'll know you've bitten off more than you can chew when you give up halfway through building a robot. In that case, step back and don't be afraid to start over and build an easier robot and then come back to the more challenging robot when you're ready.

With basic kits, assembling a robot is similar to assembling a model airplane: You just follow step-by-step instructions, putting the little plastic part A into part B and so on until, voila, you have a finished product. All you may have to do when the construction is complete is insert some batteries and turn the power on.

Don't be deceived into thinking that these robots are just expensive toys. Many are sophisticated and introduce you to essential robotic building principles.

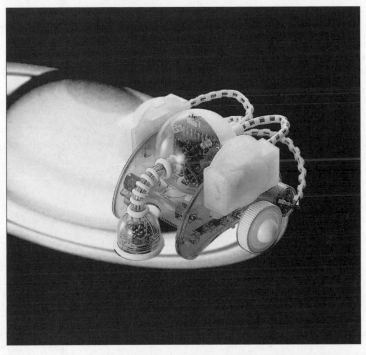

Figure 2-2:
OWI Kit's
Hyper Line
Tracker.

Although nonprogrammable kits are simple, they still require basic skills. You may be required to solder parts onto a circuit board, connect wires, and test connections. You may also be required to have some basic building skills such as assembling plastic gearboxes, gluing plastic parts, and bolting parts together.

Before you begin, you should review the kit and assembly instructions to be sure that the kit is something you can handle. If you plan to give kits to your kids, you should definitely review the instructions and perhaps even build the robot yourself first. In a few cases, you may find that even a nonprogrammable kit may be more complicated than you expected.

Remotely-operated vehicles

Most nonprogrammable robots fall into two categories: remotely-operated vehicles and preprogrammed robots. Essentially, *remotely-operated vehicles (ROVs)* require human intervention to operate them, and *preprogrammed robots* don't.

The most basic nonprogrammable robot is the remotely-operated vehicle. The type of vehicle might be controlled by radio signals, a wired tether, or some other means of remote signaling. The Soccer Jr. robot, which you build in Chapter 3, is also an ROV.

An ROV such as a radio-controlled (referred to by people in the know as simply *RC*) car may be robotic in nature, but it is not autonomous (meaning it requires human interaction to do what it does). Because these types of vehicles can't operate on their own, some robot aficionados are reluctant to call them robots at all and instead refer to them as *parabots*.

Some examples of ROVs follow:

- ✔ Radio-controlled battle robots such as those seen on television shows such as *Battlebots* and *Robot Wars*
- ✔ Unmanned subs such as those sent to search the Titanic at the bottom of the ocean
- ✔ Surgical telepresence systems that enable doctors to direct surgery on patients thousands of miles away

Preprogrammed robots

The other type of nonprogrammable robot is the preprogrammed robot. Preprogrammed robots are usually autonomous; that is, they require little or no human interaction for them to perform a task.

Many preprogrammed robots have a one-track mind. You turn them on and they do one thing, such as responding to a sound or following a line. These robots have one simple task or behavior that they carry out through hard-wired electronic circuits or preloaded computer software. Basically, the designer of a preprogrammed robot makes a decision to not allow the user to modify the behavior of the robot. This decision simplifies the robot's design, as with the Comet robot from OWI, shown in Figure 2-3, which simply responds to sounds that make it move. With other pre-programmed models, you are allowed to program limited additional functionality.

Simply be creative

The ARobot kit you'll work with in Parts II and III is an example of a programmable kit robot. The sample code fragments you'll find elsewhere in this book are provided to help you see the benefit of creating a modifiable robot. Although you don't have to know programming to complete the projects in this book, I strongly recommend that you attempt to understand the essentials of Basic programming if you want to advance in robotics.

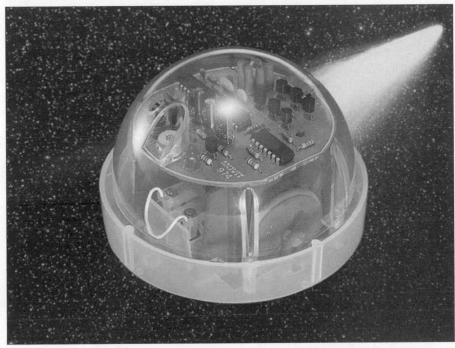

Figure 2-3:
The pre-
programmed
Comet robot
from OWI
responds
to sound.

Programmable kits

People seem to thrive on trying to change each other's behavior. A salesperson tries to get a customer to make a purchase, an Army sergeant drills the
troops to be obedient, one spouse tries to get the other to change his or her
errant ways — we even train dogs to make them fetch the newspaper or sit
up and beg.

If you're one of those people who likes to play around with the behavior of
others, you'll be happy to know that you can easily change a robot's behavior. In fact, the biggest thrill in building a robot may be getting it to do what
you want it to. Unlike nonprogrammable kits that have only one static behavior, *programmable kits* give you the ability to modify or change behavior and
create a sort of ever-changing life form. An example of a programmable kit
robot is shown in Figure 2-4. The designers of programmable robots will
never fully know what their robots are capable of because you can make
endless variations by simply modifying the software.

Getting a program into your robot

As the saying goes, there's more than one way to skin a cat — and more
than one way to program a robot. With one simple programmable robotic

kit, for example, you use a number two pencil to fill in dots on a data card that gives a list of commands (such as turn left, turn right, or go straight). Another robot kit allows you to enter a few commands into an onboard keypad.

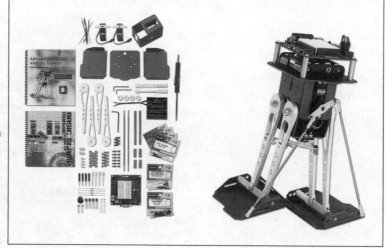

Figure 2-4:
The pro-
grammable
Parallax
Toddler
robot.

More sophisticated robots are tethered to a computer by cable or a radio link. With this kind of setup, a robot can receive commands in real time from the computer. Even more sophisticated programmable robots allow you to develop sophisticated software programs on a PC and download the programs to the robot's onboard computer brain (called a *controller*) through a data cable. The data cable download method is the most common for today's programmable robots.

Programming languages

The more sophisticated your computer, the more programming options you're likely to have. Many simple microcontrollers allow you to use only one programming language, such as Basic or Assembly. On the other hand, some robot brains are like a PC computer on wheels.

A *microcontroller,* or *controller,* contains a program, memory, and input and outputs all in a one-chip package. Small personal robots mostly use microcontrollers such as 8051, HC11, or PIC for their robot brain. The ARobot uses a Basic Stamp, which utilizes a PIC microcontroller chip. A newer term is *embedded,* which loosely means a small computer board for small devices. Embedded systems are sometimes thought of as more powerful than just a simple microcontroller board and many can run operating systems such as

DOS, Windows, or Linux. Many more sophisticated robots are beginning to use embedded system boards.

With a high-powered onboard computer, you may be able to program in quite a few software languages and interface with a few different operating systems. Such computer solutions are usually costly and complex and simply unnecessary for a basic robot. For most personal robots, a microcontroller that can run the Basic language, such as the Parallax Basic Stamp 2 controller shown in Figure 2-5, is usually sufficient to perform the common functions such as navigation and sensing the environment.

In Chapter 6, I explain some Basic language code fragments you can use with the Basic Stamp 2 controller used with the ARobot kit.

When you know how to program, you'll be well on your way to making robots that do almost anything you want them to do. Great sources for programmable robot kits include Parallax, Inc. (`www.parallaxinc.com`) and the Robot Store (`www.robotstore.com`).

Demo and sample programs

Many programmable robots come complete with one or more sample programs. The manufacturer will often give you the _source code_ so that you can simply copy the program into your program-editing software. Having the source code is a big advantage because you can simply make any needed modifications, rather than start from scratch.

When choosing a preprogrammed robot, look for a kit that has a few sample or demo programs to save you from having to develop the entire robot's program from scratch.

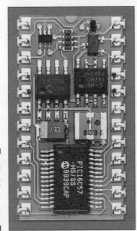

Figure 2-5:
Parallax
Basic
Stamp 2.

Moving Up to Robot Construction Sets

After you've built a few kits, you're likely to get the urge to design your own robot. Before you attempt to build a robot from scratch, you might want to check out construction sets.

Construction sets are a great way to get started with building and designing your own robots. With a good construction set, you can start by building a robot in a standard way, thus preventing unexpected design problems, and then later go off in your own design direction when you're ready to.

Anatomy of a robot construction set

You may be familiar with common hardware construction sets. The addition of a computer brain makes a construction set such as LEGO MindStorms a true robot construction set. To truly utilize this type of kit, you need to have a PC. The kit includes an *integrated development environment (IDE),* which you load into your PC to help you compose the robot's software control program, as well as a download cable for downloading programs to your robot. The advantage of computer control adds a new dimension to construction sets and allows you to build a truly autonomous robot.

One advantage of construction sets is that if you don't like the robot you've built, you can disassemble it and rebuild it another way. Of course, what can sometimes be an advantage can also be a disadvantage because the assembly isn't permanent.

With some construction sets, parts may not stay assembled as you would like and the robot may fall apart at an inopportune time. To remedy this, you can tape, rubber band, bolt, or even glue your parts together to avoid the inevitable hardware breakdown. However, use glue only when you feel you won't ever want to disassemble your robot.

With prefabricated sensor and actuator modules and a prefabricated programmable central robot brain, robot construction sets such as LEGO MindStorms (www.legomindstorms.com) and Fischertechnik (www.fischertechnik.de) have all the parts necessary to construct a robot.

Kits like these have proprietary connectors, so you can connect only their devices, such as motors, lights, and IR sensors. However, even though you're limited as to what you can connect to the proprietary robot brain, a construction set does provide many great options. You can create a robot that can follow a line, detect and follow a wall, sort cards, or even compete in a robot contest.

Grow, grow, and outgrow!

I find most people who like robots are tinkerers, and therefore tend to play around and experiment with stuff. Although robot construction sets typically include hardware and software instructions, you're not limited to one robot design. You can get started with these kits by building the sample robots, and then modifying the software or hardware to change the robot as you want.

Modifying a sample robot is a good way to try out new designs. Robot construction sets, unlike simple hardware construction sets, allow you to prototype the hardware, electronics, and software design of a robot. Hardware construction sets can also be useful to prototype a robot, but they usually don't have electronic or software facilities to help with everything a robot needs to be a robot.

After you understand the fundamentals of hardware and software construction, you may want to make the move to building a robot without the sample information provided in a basic robot kit. You may also want to purchase add-on packs for these kits. For example, you might purchase a camera add-on pack to enable robot surveillance. At some point, though, you may outgrow the limits of a construction set, which is when you might start to think about building one from scratch.

Building a Robot from Scratch

There will probably come a day when a robot kit or construction set just won't fit the bill for the task you want to accomplish. Using proprietary parts that you can find in construction sets can be both limiting and expensive. Building a robot from scratch would logically seem to be the next step.

Choosing what to get off the shelf

Because you'll always use some preassembled parts, such as a motor, a robot is never completely made from scratch. But at some point, you'll want to forego kits and construction sets and move on to a freer and less limited design approach.

Reinventing the wheel has never been my favorite pastime. If you're like me, you figure out just what parts you want to build and what pre-assembled parts will do just fine, thank you. If you can find an off-the-shelf component that is not too expensive and fits the job, such as suitable lawnmower wheels, why build it?

You can find robotic components from various robotics and catalog sources such as the Robot Store (www.robotstore.com), Parallax (www.parallaxinc.com), Acroname (www.acroname.com), Jameco (www.jameco.com), and JDR (www.jdr.com).

Here are a few approaches to building from scratch:

- **Starting from halfway there.** Sometimes building from scratch means starting with a suitable robotic base and *augmenting,* or adding to, and modifying that base to provide new features. Many robot builders start with a toy tank or truck as a base and then "hack" in a controller computer brain and a few sensors to make their dream robot. Utilizing bases that are not too difficult to hack into is a great way to make a robot from scratch. You may also want to use some parts from a kit or construction set, as well as some custom parts.

  The media has changed the meaning of the word *hacker* so that it now means something bad. The term they should be using is *computer cracker.* A *cracker* is someone who breaks into a safe or secure computer system and steals. A *hacker,* on the other hand, is someone who just hacks away at software on their computer or makes modifications to hardware so that something that wasn't meant to fit now does fit. So when I say "hack in a controller computer brain," I'm talking about a person making drastic modifications to a robot just to make the robot brain fit in where it wasn't designed to fit in. Such a thing is possible, but it requires some complex modifications (hacking) to get it to work.

- **All the way from scratch.** Unlike the preceding method, in which you add stuff to a prefab base, this construction method requires that you plan, design, and know what you want to build and how you want to build it (screws or glue, metal or wood) before you begin construction. Typically, you start from the ground up. First you decide on the method of locomotion and choose motor-driven wheels or servo-driven legs. Then you give the robot some kind of body, frame, or base. Next you add sensors and other peripherals such as cameras and grippers. You then have to develop and test the software. This is an advanced building method that requires a lot of discipline and attention to detail and a well-thought-out plan.

- **Software-only (virtual) approach.** At the other extreme, you could buy a robot that's already made. But, you say, buying a premade robot is not really building a robot from scratch. Don't forget that creating, developing, and revising software can be the biggest and most time-consuming part of your robot project. You may want to buy an already built or partially built robot, put it together, and then spend the rest of your time working on the software.

  Creating a *virtual robot,* or a robot that exists and operates only inside a computer, is not the same as creating a real robot. The real world has many variables that a virtual robot may not face, so it's hard to call a virtual robot a real robot. Simulating a robot virtually, though, may be a good way to speed the development and testing of a real robot. Often complex software algorithms are simulated in this way. When you then construct the robot in the real world, however, you may find that it doesn't cope the same as it did in the virtual world.

✔ **Starting with a prebuilt robot.** Some robot builders don't want to be bothered with developing hardware; they're interested only in creating the software for their robot. Having a prebuilt robot allows you to forego building the hardware and thrusts you right into building the software. This method is different than building a virtual robot because your robot will be facing the real world. Building a robot's hardware usually takes only a short time, but creating and tweaking the software can take a lifetime. Don't worry, most prebuilt robots allow plenty of room for hardware expansion, such as adding new sensors or more powerful controllers.

Parts sources

Radio Shack (www.radioshack.com) is one of the most popular sources for electronics parts. With over 7000 stores nationwide, the convenience of patronizing this retailer simply can't be ignored. However, you'll pay for that convenience. In this section, I tell you about some less expensive parts sources for you to check out.

Come and get your free catalogs!

To build a parts source library, head straight to the Internet or the post office to send your requests for catalogs to electronic and hardware parts warehouses. Most catalogs are free (and the ones that aren't probably aren't worth having).

Check out the advertising section of electronics and hardware magazines for ads from companies that offer free catalogs. You may have to wait patiently by the mailbox, though, because some catalogs are sent out only four times a year.

Some great electronics and parts catalog companies include Radio Shack (www.radioshack.com), Mouser (www.mouser.com), Marlin P. Jones (www.mpja.com), Jameco (www.jameco.com), and JDR (www.jdr.com).

Scrounging for parts

Just because you live online, don't ignore local resources. Almost every large city has one or more electronics or hardware businesses. Scan the yellow pages to find such sources in your area. Also, almost every large city has a scrap metal store.

Dumpster diving

Some manufacturer's dumpsters may hold a gold mine of throwaway electronic parts. However, if you take my advice, you'll save time, money, and potential frustration by going to a reputable supplier.

That said, be careful when you go to your local electronics swap meet. You may find that some of the treasures that you thought you were getting are dead hardware retrieved from a dumpster.

In addition, don't fill up your garage with parts just for the sake of filling up your garage. Have a need in mind, and don't buy more than you need. It's better to have to buy an overpriced electronic part once in awhile than to stockpile a bunch of parts you'll never use.

Some of my best buys have come from buddies at monthly computer or radio swap meets wanting to get rid of their load of trash, which turned out to be my load of treasure. Browse these locations with money in hand and develop good negotiating skills to help add to your bounty.

Economics and Time Considerations

Building robots is fun and rewarding, but everything comes at a cost. In the case of robotics, it all boils down to time and money. Robot building is sort of like going to college: You'll discover a lot, but you must devote a lot of time, energy, and money to get anything out of it.

To avoid going bankrupt building robots, take a look at the following methods for helping you avoid the black holes of time and money.

Start small

First, evaluate your time and money limitations. Besides the robot itself, you need to purchase special tools, software, sensors, and other hardware. Hardware will not be too costly and will take only a little of your time to build, but creating software can easily take all your spare time and then some.

It's possible to build robots on a budget. Start with a simple, inexpensive kit for now and work up to costlier, more sophisticated robotics — with all the hardware and software bells and whistles — when you get that next raise.

Haste makes waste

You'll need to figure out how to build robots inexpensively and efficiently if you plan to stay in robotics, but don't cut corners. Don't try to build a quickie, cheap robot with hot glue and cardboard. Sometimes spending a few extra bucks on a quality item can make a huge difference in the longevity and performance of your robot.

Also, don't rush things. Usually, you won't be in a race to get your robot finished on a deadline, so stop and give each step a little thought. Read all directions carefully and know what you're getting into before you put a robotic kit together. Taking your time means you won't make costly mistakes.

Don't get stuck in a rut!

Your robot will never be perfect, so you'll need to decide when to stop and be happy with what you have. Don't get bogged down in fine tuning something on your robot. If you stop enjoying one robot endeavor, move on to the next project.

If you get stuck with a bad part, decide whether you should cut your losses. Sometimes a contrary actuator or sensor isn't worth fooling with; just go buy another one.

Reuse and recycle

Writing and modifying software can eat up a lot of your time. Try to reuse software functions and libraries you've already created or use software you can get from the Internet.

To be a good robotics dude, you may want to offer your own code to others on the Internet as well. Your generosity will be repaid when an online buddy helps you out down the road.

If you're having trouble debugging a piece of software, don't spend too much time before asking for help from a friend. You may also want to purchase or acquire a software debugger to help with the time-consuming and frustrating task of finding errors. Write software carefully the first time, and you'll have less need of debugging later.

You can extend the reuse-and-recycle concept to hardware design too. After all, you may become a mentor to others, and they can benefit from your designs, ideas, and mistakes. So keep a notebook of your designs that includes dimensions and drawings so others can build on your work.

Get testy with your robot

Carefully testing your robot at each step can save you time. You should test each part (*unit tests*) and also test the robot as a whole after it's fully assembled (called an *integration test*). If you don't test the small single pieces before you assemble your robot, it will be much harder to determine what has failed or is malfunctioning when you get it all together. If you add or change part of your robot, retest everything (called *regression testing*) to be sure other parts of your robot were not affected adversely.

Never underestimate the value of testing. If you test your work up front, you'll have less testing to do after you finish.

Get on the right path

The time will come when you'll need to decide whether you want to start with a robot kit, a construction set, or from scratch. Choose your robot construction path wisely, my friend, for the days are short and the money is shorter. Keep your eyes squarely on your goal and remember that the important thing is to have fun. Otherwise, what's the point?

Chapter 3

Building Your First Robot

*W*ouldn't it be nice to start with a kit that you know would work? The startup time is minimal, the cost is manageable, and the required tools are reasonable. Within a few hours, you could have a working robot, and although it might not be the most powerful robot on the planet, it would make a good springboard for your next project.

If you're tired of seeing others having all the fun, it's time to dig in and build your own robot.

Robot Kit to the Rescue

Putting together a kit can familiarize you with robot terminology and components. And the process might give you ideas about what you want out of your next robot. After all, building robots is a constantly expanding endeavor.

After looking through many catalogs and wading through online data sheets, I selected the Soccer Jr. robot from OWI, Inc. (www.owirobot.com or call 310-555-6800) as the kit to use in this chapter. This little plastic robot not only performs a task but also offers some human interaction and control.

The kit comes with a wired controller that allows it to move in any direction as well as capture and shoot ping-pong balls disguised as small soccer balls. Because a human can control Soccer Jr., it's great for competitions too. Figure 3-1 shows the little fella after assembly.

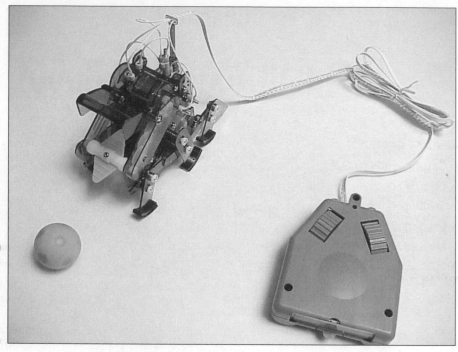

Figure 3-1:
Soccer Jr.,
a non-
programma-
ble robot.

On Your Mark, Get Set . . .

After you purchase Soccer Jr., you need a few more things before you begin your robot building adventure:

- About two hours of your time
- A few hand tools: screwdriver, pliers, cutters, hammer, and knife
- A place to work (a table or the floor)

Making time

How long this project takes to complete varies by how fast you are, how many times you're interrupted, and any trouble you may encounter. I suggest beginning only when you can commit at least one hour. It's not practical to work on a kit like this in five-minute increments. My favorite time to work is late at night when the phone isn't ringing and interruptions are minimal.

Your workspace

Most people use the kitchen table to build models and kits, but doing so has some disadvantages. The kitchen table is usually in a high-traffic area of the house and accessible to youngsters. Cutting or burning the surface may be frowned upon by other members of your family, and eventually the table will be needed for its intended purpose, requiring you to move elsewhere.

Another option is a small card table. It's easy to set up and move, so you can put it in a back room away from the commotion of daily life. Some card tables have a soft surface that conveniently prevents parts from moving around.

My favorite workspace option for building robots is a carpeted floor in a back room away from traffic. The parts don't move around and get lost and, frankly, I find it more comfortable.

You'll want to protect your work surface from things such as knife cuts and glue. Use a wooden board, a piece of cardboard, or a lunch tray to avoid little robot-building disasters.

Tools and grunting noises

Using and collecting tools is definitely the second most fun part of building robots — second only to playing with them. Most robot builders, like cabinet-makers and artists, become attached to their tools and usually have favorites. There's nothing like having the perfect tool for the job.

For Soccer Jr., you need only a few common hand tools such as a hammer, a screwdriver, a hobby knife, and pliers, as shown in Figure 3-2. Usually, the instructions for the kit you choose will indicate which tools you need.

If you spend the time to care for your tools, they should last a lifetime. Pick out a tool box the right size for the quantity and size of your tools and keep them in there. If others borrow your tools, make sure they return them or you'll end up spending ten minutes searching just to do a simple project. If there are others with tools around you, use colored electrical tape or etch your name in your tools to identify them as yours.

Finally, keep your tools away from moisture. Rust can ruin an expensive pair of wire cutters and make your screwdrivers yucky. It may be necessary to keep a light glaze of oil on your tools if they are exposed to high humidity. This is especially true of drills and saws that have large, exposed metal surfaces.

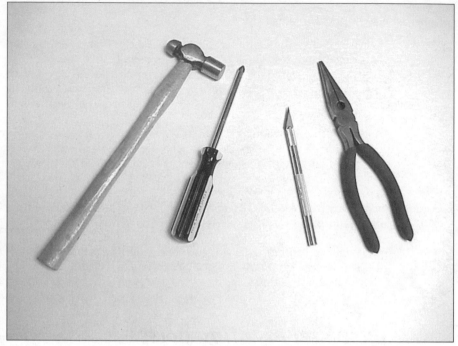

Figure 3-2:
Building a
robot is
impossible
without the
right tools.

Precautions

One of the first parts of the assembly book is the precaution section. I suggest you read it, but I'd like to highlight the items that, in my opinion, deserve the most attention.

Using a hobby knife

Like most robot kits, Soccer Jr. is made of plastic. Plastic parts are usually connected in the molding process and need to be separated before you use them. And after you separate the parts, you may have to trim them to get rid of those jagged plastic edges. The tool for these jobs is a standard hobby knife.

Most hobby knives have a metal handle and a removable blade at the end. The blades are extremely sharp and can easily cut human skin. The obvious danger is that you might slip and accidentally cut your finger or other valuable body part.

Be wise and always cut away from your body. Form this habit early in your robot-building career: It could save you a few pints of blood and boxes of bandages.

When you need to apply a large force while cutting, set the part on your work surface (preferably on an old piece of wood or cardboard), and cut down toward the table with your fingers above the blade. Figure 3-3 shows this technique in action.

Another problem with hobby knives is that they're usually cylindrical and will easily roll, sometimes right off your work surface. In fact, it's a well-known fact that hobby knives have the uncanny knack to land with the blade pointing down. As you might imagine, this puts your legs and feet at risk, not to mention passing pets and small children.

There are several solutions to the danger of flying hobby knives. You can use the plastic cap that covers the blade, stab the blade into a chunk of Styrofoam, or purchase a knife that has a noncylindrical rubber grip that prevents it from rolling willy-nilly.

The bite of pliers

A common problem when using pliers is when they slip, pinching your finger. This problem also exists with needle-nose pliers, wire cutters, and wire strippers. It hurts, and if you haven't experienced it, it's only a matter of time.

The simple solution is to make sure your fingers are not between the handles near the joint of the pliers. Sadly, most of us have to learn this the hard way.

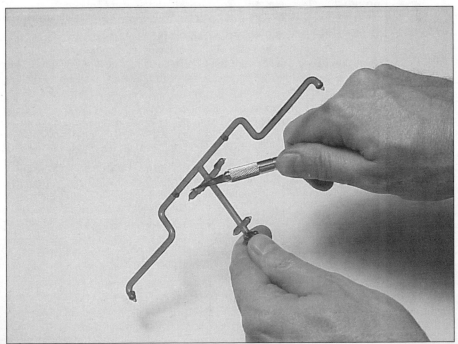

Figure 3-3:
Always cut away from your body.

Problems with small parts

Even a simple kit like Soccer Jr. can have a hundred or more small parts, including screws, washers, and gears. There are two main problems with small parts: They're easily lost (usually by falling off the table and rolling halfway to Siberia), and they're easily swallowed by little kids, or even pets.

The solutions are obvious and easy. Use small bowls to hold parts, especially small metal hardware, and make sure the younger ones are busy watching cartoons in another room, preferably separated by a door or two.

It's in the book

It's virtually impossible to put together a kit of this complexity without using the book, so resist the urge to build freestyle. Even though it may be obvious where a component attaches, there may be a special sequence required to make everything fit together. Take the time to go through the instructions in order and you'll be rewarded with less frustration and a quicker build time.

Let the Building Begin

Enough chitchat, let's get down to business. Start by removing all the parts from the box. Lay them out and open the assembly book, but don't remove the parts from their bags yet.

I won't be going over each step in the construction of Soccer Jr. because the manual does a fine job. However, I will discuss a few areas that you might struggle with or have questions about.

If you're overwhelmed by the amount of parts, don't fear. Building a robot is like washing an elephant — just do it one small area at a time. The designers of the kit have packaged the parts in groups. You'll also find that the drawings in the book give you all the information needed to assemble the robot, so if something doesn't make sense, study the drawings.

Remove the plastic cover for the gears and hardware but leave the parts inside the convenient trays. Remove the plastic parts from the bags but keep them in separate groups. Check the bags to make sure a small part hasn't separated from the others.

Jumping ahead

Let me start by encouraging you to resist the urge to jump ahead on the assembly process. If you do jump ahead, the building process will take you longer, I promise. You must follow a certain sequence to get everything to fit together.

Going nuts

The Soccer Jr. kit has two types of nuts: a plain hex nut and a lock nut. You've probably seen plain hex nuts, but you might not be familiar with lock nuts. Lock nuts are special nuts that are taller than normal nuts and have an internal plastic ring to help the nut resist vibration. Lock nuts are used in applications in which motion and vibration are present and when using a lock washer is impractical. Lock nuts are harder to install than normal nuts and usually require a nut driver (a screwdriver-like tool that handles nuts) or pliers.

Cutting up

The plastic parts are attached together as part of the molding process. You'll be detaching them using your hobby knife or wire cutters. Resist the urge to clip them all out at once. Instead, wait until the part is called for in the assembly instructions.

The story of gears and motors

Three motors are used in the Soccer Jr. kit, and each one requires a small pinion gear. The gear must be friction fit onto the motor's shaft, and the force required is too great to do with your fingers. The best method I've found is to place the gear on a flat surface, place the motor shaft into the gear's hole, and then lightly tap the motor's shaft that's protruding from the rear. You can use a small hammer or the flat side of a pair of pliers. This method is shown in Figure 3-4. Do not apply force to the motor's body or the shaft will actually move out of the motor.

Batteries are last

If you install the batteries before you finish the assembly, the wires will have voltage on them and you could create a short, causing the batteries to drain. So, when you feel the desire to install the batteries, resist! Install them only when the instructions call for them.

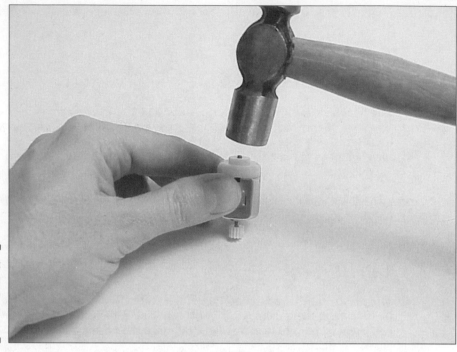

Figure 3-4:
Installing the pinion gears onto the motors.

Extra parts

When you're finished, you'll notice some extra parts such as screws, nuts, and some plastic parts. Don't panic. The manufacturer puts extra parts in there just in case you lose one or they make a mistake. I suggest that you put all the extra parts in a bag and keep them for a while just in case you need one of them later. You could also save any extra hardware for your own robot parts inventory.

Testing and Troubleshooting

You've endured a multitude of plastic parts and metal screws, and now you have a nice little robot to show for it. Although Soccer Jr. is relatively simple, things can still go wrong, so I'll deal with a few possible problems here.

The hand controller, which is shown in Figure 3-5, has three 2-position switches. At the top, rocker-style switches control the direction of the soccer ball motor. Press on each side to see that it's operating properly — the rubber flap should spin.

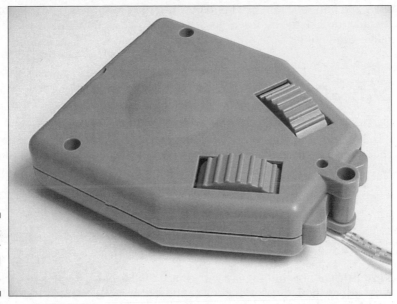

Figure 3-5:
Soccer
Jr.'s wired
controller.

The other two switches are slide-style switches that make the leg motors move. Each switch has two positions: one for forward and one for reverse. Test them to make sure all is well. Pressing both buttons in the forward direction should make the robot move forward in a relatively straight line. Pressing only one switch should make the robot turn.

The dead robot

The easiest type of problem to solve with electronic or computer equipment is when it's completely dead. Power is usually the culprit in those situations. In Soccer Jr.'s case, power is supplied by four batteries. It's important that these batteries are installed correctly, so make sure they match up with the indicators on the battery holder.

Motor problems

Miswiring a motor will cause it to go in the opposite direction or not move at all. Make careful note of the drawing that shows where each wire goes and make sure that the exposed metal part of each wire is touching the screw post that transfers power to the motor.

Also, did you remember to bend out the small terminals on the motors before installation like the instructions showed? You do that to allow the screws to make good contact.

Side stepping

The mechanical linkages that make the six legs operate can be a bit confusing. If the legs are moving correctly, double-check the linkage and the hardware that connects them. The assembly guide has a few good pictures that will help. If the motor is spinning but the legs are not moving at all, make sure that the lock nut on the leg cam is tight.

Soccer Jr. in Action

This little robot is fun to play with, but it takes some practice to make it do what you want. I found it difficult to move the robot towards the ball and shoot the ball, but after a short while I got pretty good at it.

One of the reasons I chose Soccer Jr. for this project was that it involves user interaction. So many nonprogrammable robots on the market do one task without operator intervention, and people tend to lose interest quickly. If you spend two hours building a robot, you should expect at least two hours of fun! Soccer Jr. does a fine job of keeping my attention. It would be even better if there were two of them so that people could play against each other.

Having two robots makes competitions fun because both robots can go against each other at the same time, but competitions are certainly possible with only one robot. The user guide gives a few suggestions, which I list next, and I've added a few of my own.

Soccer

It should come as no surprise that Soccer Jr. is perfect for soccer, as Figure 3-6 shows. Start with a field the size of a ping-pong table. Put a rim around the table or area to prevent the balls from shooting out. You can use plastic piping, strips of wood molding, or cardboard. On each end of the field, mark an area for the goal. Be sure the size is large enough, or you'll get discouraged. Make a line across the center using yarn or masking tape.

You can make up any rules you like but the obvious object of the game is to shoot the ball into the goal. It's more fun with two robots, but you can also play with just one. Use a kitchen timer to limit the time each player has to make a goal. When a goal is made, move the robot past the middle line and place the ball in a location that requires work to reach. Whoever wins gets to take the author to dinner.

Gathering balls

The suggestion of gathering balls is made in the manual and calls for the players to collect several balls and shoot them into a goal. A time limit is set and players take turns. Ping-pong balls can be used as additional balls. The book suggests cutting a hole in the balls and inserting small pieces of rubber bands to prevent the balls from rolling so far.

Shooting pool

The goal of the next game is to shoot six balls that you place in the center of a pool table into the pockets. The robot has rubber feet, so it shouldn't damage the surface of the table.

Extra points are given for the player who puts a ball in each pocket. You could also impose a time limit using a kitchen timer. Two minutes should be a short enough time to stress out even a seasoned robot master!

Figure 3-6:
Soccer Jr.
on the go.

Sumo without the bulk

A sumo-wrestling contest is possible only with two or more robots. Make a circle or square that's about three feet across. Place the robots inside the area and start the match. The goal is to push your opponents out of bounds. The last robot left is the sumo champion.

Learning Your Robot Chops

You might think you didn't discover much through this adventure. But you did realize the value of following directions, you found out the proper way to use a hobby knife, and you gained skills for coping with the stress of shooting a bunch of balls with a time limitation. Now I'll talk a bit about some technical concepts used in a simple robot like Soccer Jr.

Motors making my head spin

Soccer Jr. has three small DC motors. They're called *DC* because they use direct current, meaning the voltage stays at a steady polarity and doesn't switch back and forth as in AC (alternating current). A familiar source of direct current is a battery.

DC motors like the one in Figure 3-7 are common in toys and other small products that run on batteries and perform mechanical work. Battery-powered cars, trains, and electric cranes sometimes have many motors, all controlled manually or with a small computer controller. Even pager vibrators are just small DC motors with an off-centered weight.

DC motors have two terminals to accept power. The polarity (positive or negative) of the voltage applied to the terminals determines the direction that the motor spins. (You've dealt with polarity when you took care to install a battery with the + end and – end positioned correctly in a battery slot.)

In the Soccer Jr. robot, each motor's direction is controlled by a three-position switch. In the center position, no power is applied to the motor. To each side of the center, opposite polarities of voltage are applied to control the motor's direction.

Did you know that DC motors can also work in reverse to create electricity from motion? When used like that, they're called generators. This is the way your car's engine creates 12 volts to run the radio, lights, and spark plugs. Generators can also be found in modern windmills, which create power from the wind. The polarity of the voltage produced is determined by the direction in which the motor spins.

Grinding those gears

DC motors typically spin fast and have little torque (strength). Most things you'll ask your robot to do need a slower, stronger motion — that's where gears come in. With a properly configured gear train, you can reduce the speed and increase the torque — or the other way around if needed.

Here's how this works: If you place a gear with a small diameter on a motor shaft and let it mesh with a gear that has a larger diameter, the speed of the larger gear is slower than that of the motor. Figure 3-8 shows the action of a pair of gears. The reverse is also true: A large gear on a motor meshing with a smaller gear increases the speed.

The amount of change in the speed is a function of the ratio of the gears. For example, if the smaller gear is ¼ the diameter of the larger gear, the speed is decreased by a factor of 4, and the ratio is referred to as 1:4.

To get greater speed reductions, you simply place several pairs of gears (called *stages*) in series. Each stage has its own ratio, and all of the stage's ratios multiply. The same is true of gear configurations that increase speed.

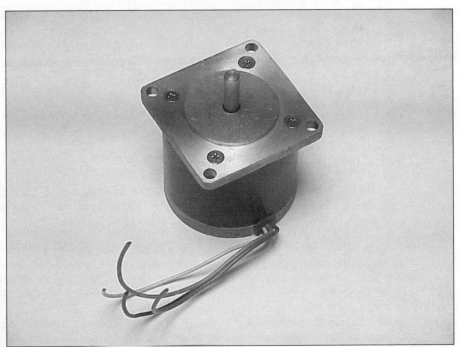

Figure 3-7:
A DC gear.

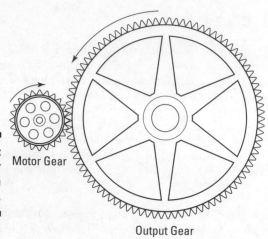

Figure 3-8:
Gear
reduction
in action.

Motor Gear

Output Gear

When you change the speed using gears, the torque changes as well. If you increase the speed, you reduce the torque by the same ratio. And if you decrease the speed, you increase the torque. Some power is lost just driving the gears, so the overall power is somewhat less than ideal.

Belts and pulleys are used in the design of some robots to change the speed and torque in the same way gears do. The change is a function of the diameter of the pulleys involved. Belts and pulleys are quieter, are usually cheaper, and can transmit power a greater distance than gears. You can find belt-and-pulley systems in your automobile, on drill presses, and in air-conditioning systems. The chain and sprocket on a bicycle work on the same principal as belts and pulleys.

Cams, but not for the Web

Pairs of gears in a motor produce a circular, or rotary, motion. If you wanted your robot to spin in circles all day, that would be fine. But if you want your robot friend to move in a straight line, you must change that circular impulse to a straight one.

Cams are interesting little devices used to convert rotary motion to linear, back-and-forth motion, as shown in Figure 3-9. Soccer Jr. uses a cam to move its feet, for example. As the cam turns, the linkage follows the cam's wheel and produces a cyclic linear motion that makes your robot walk the straight and narrow. (You might have seen this type of action on the wheels of a steam engine. But in that case, the engine pushes the rod, which turns the wheels.)

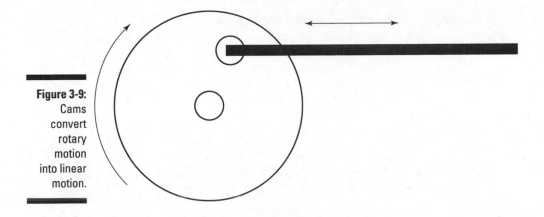

Figure 3-9:
Cams
convert
rotary
motion
into linear
motion.

Juicing it up with batteries

In this section, I discuss a few battery options such as the ones shown in Figure 3-10 — including those so-called "terribly expensive" rechargeable types.

Figure 3-10:
Battery
options
galore.

Those batteries you buy at the grocery store

Little pink bunnies marching around your TV screen aside, alkaline batteries have a shorter lifetime than a politician's promises. These batteries are everywhere, and everyone has a junk drawer full of them. Why do we buy these relatively short-lived beauties? The main reason is that they're initially cheaper than rechargeable types and don't require an expensive charger.

An alkaline battery will often last longer and have a higher voltage than a single cycle of a rechargeable battery, but after the charge is gone, that's it. For that reason, I'm about to put the notion of alkaline batteries being a better value to sleep for good by telling you all about the benefits of rechargeable batteries.

Regardless of my preference for rechargeable batteries, alkaline batteries will work just fine in your robot.

NiCd batteries

A dying technology called NiCd, or nickel cadmium, offers rechargeable technology but has some downsides. They're bad news for landfills and also suffer from something called the memory effect. The memory effect causes a problem if you charge the battery before it's completely discharged. Essentially, the battery will charge only a percentage of its total capacity and eventually loses the capability to fully charge. Bottom line: I suggest you don't even consider NiCd batteries for your robot.

NiMH batteries

A newer rechargeable battery technology is nickel metal hydride (NiMH). This type of battery has greater capacity than the older NiCd battery, and it doesn't suffer from memory effect.

The main disadvantages to NiMH batteries are that they're expensive and require a charger made specifically for them. But a careful look reveals the truth behind the cost-value tradeoff for this type of battery.

Consider this: ARobot requires eight AA batteries. This costs about $6 for nonrechargeable alkaline batteries or about $25 for NiMH rechargeable batteries, plus another $25 for a charger.

Rechargeables sound expensive, huh? Not so. In this scenario, if you replace the alkaline batteries ten times, you've spent $60. Buying NiMH batteries and the charger would cost you $50 — and your batteries would still have hundreds of recharge cycles left. After 100 cycles, alkaline cells would cost $600, and NiMH batteries would still have put you out only $50.

These calculations disregard the negligible cost of electricity for charging, but trust me, you're still way ahead with rechargeables.

You can easily see how the so-called cost advantage of cheaper alkaline batteries is no advantage at all. For ARobot, you should definitely invest in NiMHs and a charger. (You can see what one looks like in Figure 3-11.) The only question left is, how do you spend all that money you're going to save?

Generally, the lighter the batteries, the less work your robot requires to move around and, therefore, the longer your battery life. NiCad batteries are usually the lightest but can be recharged only a limited number of times. NiMH batteries are heavier and can be charged many times. You can tell how much power a battery will supply by their amp-hour (Ah) rating. For example, a 7Ah battery can supply 1 amp for 7 hours or 7 amps for 1 hour or 28 amps for 15 minutes.

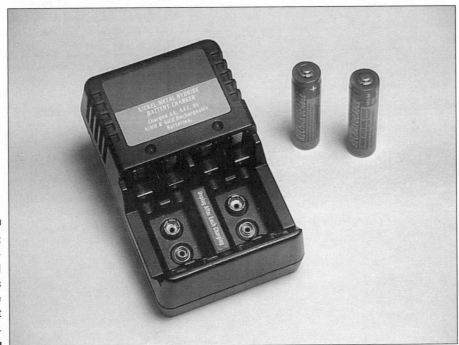

Figure 3-11:
Recharge-
able NiMH
batteries
are the
cheapest
option.

Baby Robot Steps

If Soccer Jr. was your first robot-building project, you might be a bit disappointed, especially because most expectations about robots come from the big screen. In movies, robots can do almost anything, tread any terrain — and most even have a charming personality. The harsh reality is that robots are just machines with components, like your car and personal computer. Robot technology is advancing every year, but we still won't have movie-like robots for quite some time.

Small, nonprogrammable robots like Soccer Jr. are great fun, but they show us that robots need a brain, sensors, and greater capabilities to become truly useful. The technology to do these things is still difficult and expensive to implement, but it's coming. The rest of this book delves into the more complex arena of programmable robots, complete with programs and sensors. You'll find this next step a largebut rewarding one if you're serious about robotics.

Part II
Programmable Robot Prep

The 5th Wave   By Rich Tennant

Ned Beally, of Beally Construction Co., helps his children with a robotics project.

@RICHTENNANT

"Oh, big surprise — another announcement of cost overruns and a delayed completion date."

In this part . . .

If you were baking a cake, before you could bake a thing you'd have to clear a work surface and assemble your ingredients. You'd have to turn on your oven and gather cooking tools such as a spoon, a pan, and a bowl.

Well, robot building requires some basic preparations that aren't all that different from cooking. You need an efficient workspace for building. You have to gather ingredients (the parts of a robot kit plus wire, screws, and so on) and tools (pliers, cutting knife, and more).

One difference between a cake and a robot, however (no analogy can go on forever), is that with programmable robots you can also download programs to make the robot do things. (Try that with a cake and you'll be sorely disappointed.) So the last chapter in this part is a quick overview of programming basics for the programming challenged. But don't worry — the type of programming you have to do in this book is . . . well, a piece of cake.

Chapter 4

Setting Up Your Robot Workshop

• •

• •

*T*o set up a fantasy robotic workshop, you would need millions of dollars and a crew of PhDs at your beck and call. Most of us don't have those kinds of resources, so we make do with what we can afford, which works just fine most of the time.

In Chapter 3, you built your first robot, a nonprogrammable machine called Soccer Jr. Now it's time to move on to bigger and better robots. In this chapter, you discover how to plan your robotic workshop for optimal functionality and how to set up and organize everything from shelving to your tool kit. I tell you all about the testing equipment you use to check things such as electrical charges and other scientific stuff. Then I go into the basics of soldering — a skill you must have to be a robotic whiz.

After you've read this chapter, you'll be able to assemble all the basics you need for your very own robotic workshop.

Creating an Ideal Work Area

Setting up a robot workshop is a little different than setting up just about any other kind of workshop. That's because when you build robots, you're dealing not only with garden-variety toolbox tools but also with electronics (see Figure 4-1) and perhaps computer hardware and software.

Because of this hodge-podge of equipment, your workshop may look like a mad scientist's lair. (Just be careful that the local townsfolk don't suspect you of making little Frankenstein robots!) Because robotics is all about research and development and trying things to see what works the best, don't feel like you have to please anyone but yourself with your workshop. It should be a place where you can feel creative and draw inspiration for new things to do with your robot. Don't let my recommendations for a workshop deter you from making your space exactly the way you want it to be.

You don't need a fancy shop with specialized shelving and a hermetically-sealed environment to build robots. But you do need to consider where to put your workspace, what type of work surfaces you need, and how to get organized. That's what this section is about.

When you're ready to move from simple kit assembly to full-blown, do-it-from-scratch, programmable robot building, you'll need plenty of room and a good-sized desk or long table that can accommodate a PC, electronics gear, test equipment, and a robotics work area. A lot of your time will be spent editing software and downloading that software to a robot, so be sure you have room on your desktop to work at your computer and a good desk chair to keep you comfy during hours of programming work.

Don't put your workshop in a humid, dusty old garage. A garage is great for housing large machine-shop equipment, but you — and your PC and electronics — will be much more comfortable working in a dry environment with a minimum of dust.

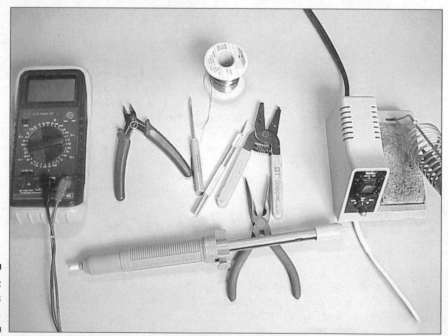

Figure 4-1:
A robotics
workshop.

In addition to a work surface, you'll need some storage space. If you don't use space efficiently, you can fill a room with tables and shelves and still find that you don't have enough storage space.

Getting Organized

Now that you're exploring the world of robots, you'll encounter more parts than will fit in a two-car garage. Even a simple kit can have a hundred or more small parts, including screws, washers, and gears. Small parts have two main problems. One, they're easily lost (usually they jump off the table and make a break for Siberia). Two, they're easily swallowed by little kids or your pet greyhound.

The solutions are obvious and easy. Use small bowls or plastic bags to hold parts, especially small metal hardware pieces. If you end up building a lot of robots, consider getting bins or drawers (see Figure 4-2 for ideas) to help you organize the parts. You might use one set of bins for resistors, another for capacitors, another for ICs, another for connectors, another for hardware such as nuts and bolts, and yet another for all those miscellaneous parts. Label each drawer or bin so you can easily find the contents hidden within.

Figure 4-2:
Bins and storage drawers.

Next, add a shelf or two to hold all those bins. If you don't keep all your parts organized, you'll end up buying duplicate parts because you can't remember where you put the ones you already bought.

Some parts won't fit into bins, such as lawn mower wheels, DC motors, sheet metal, and angle aluminum. Remember that you'll have to devote some of your shelf space to these.

Tool Time

For simple projects, you need only a few common hand tools such as a screwdriver, a hobby knife, wire cutters, and a set of pliers. More complex kits might require a soldering iron, wire strippers, a drill, and perhaps even an electric rotary tool. And for those giant robot warriors, you might need a welder, a band saw, and a mill.

Thinking ahead and having the right tools can mean the difference between finishing the job now and waiting in frustration until the hardware store opens in the morning.

Many robots require smaller tools than the ones you would use around the house or on a car. A set of small screwdrivers (such as jeweler's screwdrivers) and a set of small ratchets (ranging in size from ¾₁₆ inch to ½ inch) may be your best friends. And it's a good idea to have a magnifying glass handy to inspect those small parts.

You'll typically use your hand tools only during the building phase of your robot's life cycle, which is the shorter part of your project relative to the time you spend programming your robot. Make sure you have a good place to keep your tools in between uses.

Personal computer

Some simple robot kits don't require any special tools, but as you advance and begin to program robots or build them from scratch, you'll need a heftier tool kit that may even include a computer.

Should you buy a computer just for your robot? Probably not when you're just starting out, unless you have more spare change than I have. But without a dedicated computer, you might find yourself commandeering the family PC for robot service on a regular basis because you can use a PC for a great many robot-related tasks. For example, you can design your robot with CAD software. You can also create and download programs from your computer to your robot.

Good tool hunting

Using and collecting tools is a fun part of building robots — second only to playing with robots. Most robot builders become attached to their tools and develop favorites, just like any artisan. There's nothing like having the perfect tool for the job.

Take the time to care for your tools; then there's no reason why they shouldn't last a lifetime (which will impress you more or less depending on how old you are). Choose a toolbox that will accommodate your tools and have it handy any time you have a robot-building urge. If others tend to borrow your tools — or if household members have their own tools — use colored electrical tape or etch your name on your tools to identify them as yours.

Finally, keep your tools away from moisture. Rust can ruin an expensive pair of wire cutters and make your screwdrivers yucky. It may be necessary to keep a light glaze of oil on your tools if they're exposed to high humidity. This is especially true of drills and saws that have large, exposed metal surfaces.

You may want to use your PC also to investigate the latest robot kits on the Internet or search online catalogs for parts. In fact, the PC is probably one of the more useful tools to have in your robotic workshop.

If you become a serious robot geek, it may be time to invest in a PC just for your robot activities and keep it in your workshop. Look for a low-price PC bargain or even a used model (but not much more than a year or two old). Be sure that your PC has an RS-232 serial port for downloading programs to your robot. If your computer doesn't have an RS-232 port but does have a USB port, you may be able to purchase a USB-to-serial adapter to do the job. Also, verify that your PC is new enough that it can run the operating system and software you need to work with your robot.

Your robot and your PC may have a somewhat shocking relationship. Be sure to protect your PC from static electricity when connecting and disconnecting a data cable to your robot. You may even want to turn off your computer before attaching a data cable just to be safe. Always ground your PC and your robot before making contact. On the PC side, connecting your PC to a three-pronged outlet is sufficient. To protect the robot, you may want to touch the frame of your robot to a grounded appliance or metal desk before connecting the data cable. Don't touch exposed data pins at the end of the cable with your fingers because you may cause a static discharge and damage a data port on your robot or PC.

Testing equipment

Testing occurs throughout the life cycle of building and programming your robot. The best approach is to test lights and batteries and motors at each stage. If you leave testing until your robot is entirely built and it doesn't function the way you thought it would, backtracking to figure out where the problem is can be a mind-boggling process.

What you test depends on what you want your robot or a particular part to do. If you want to test whether your robot will stay on for five days, you should probably use a multimeter to verify current (amps) usage. If you want to test whether a chip is getting the right electronic signals, you should probably use a data analyzer. If you want to test whether your robot has memorized Asimov's robotic laws, you should probably use a multiple-choice test.

After you figure out what you want to test, it's easy to choose the right equipment to verify that you're indeed getting the answer you expected. For example, before you bolt a 12-volt motor onto your robot's chassis, you should hook up a 12-volt battery to it to see whether the motor turns, rotates in the right direction, and turns at the right speed. Before you attach blinking lights, hook them up to a battery with the appropriate voltage to see whether they blink with the expected brightness and speed. Before you attach a sensor, hook up a multimeter or an oscilloscope to find out whether the sensor provides the data you expected. Before you hook up your computer brain to the rest of the robot and watch it sit there like a lifeless rock or, worse yet, become a flailing killing machine, test the software with a debugger and test that the output signals are within the expected specifications with a logic probe or other electronic test equipment.

Multimeter

For most robotic circuit analysis, a multimeter is probably all you'll need. It's handy for measuring electrical qualities such as voltage, amperage, resistance, and conductivity. Your garden variety multimeter is shown in Figure 4-3.

Some multimeters are autoranging, which means that you don't have to worry that what you're measuring is in a particular range. Others are not autoranging, and you can damage them if you try to measure something out of a specified range.

To use a multimeter to measure electrical qualities, first select what you want to measure, such as resistance or voltage, and then touch the two wire probes to what you want to measure. The value of the measurement appears on the multimeter's numeric display.

Suppose that you want to measure the voltage of a AA battery. You would set the multimeter to read voltages less than 10 volts and then touch the black ground probe to the negative side of the battery and the red positive probe wire to the positive side of the battery. Your multimeter should now display the voltage reading of the battery. Don't be surprised if your reading is not 1.5 volts. A AA battery that's a little run down might read 1.2 volts or less. A new AA battery might read a little more than 1.6 volts.

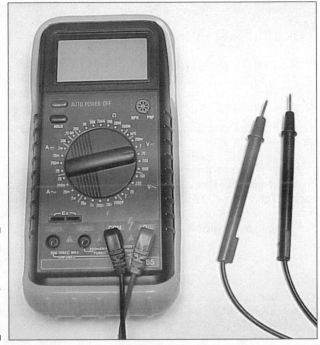

Figure 4-3:
A hand-held multi-meter for measuring electricity.

There's much more you can do with a multimeter. Consult your manufacturer's manual on how to use your multimeter to measure various electric qualities.

Be careful to use the appropriate voltage for the device you're testing. If you use the wrong voltage when testing, you may burn out the device before you ever get it on your robot. Always verify the correct electrical specifications of your devices before testing them.

Data analyzer

Data analyzer is a general term used to describe devices such as logic probes that can — logically enough — help you analyze your data. For example, you could use a logic probe to see whether data is a high voltage level or low voltage level, or you could use a serial cable and a PC to view data stored on your robot.

Essentially, data analysis is something you do to verify that your data is correct or what you expected it to be. When you download program data from a PC, for example, you should verify that the data you downloaded is correct. Many download programs have a verification mode to read back what was downloaded and compare it against original data.

Testing with a PC

A PC can be a great piece of test gear. If your robot allows serial data access, you can use a serial cable to test certain robot functions, such as sensor reading and motion control. Before some robots, such as ARobot and its Basic Stamp 2 controller, can be controlled using serial data access, you must write a special program and then download it to the robot.

If you log output data while your robot is running, you'll want to take a look at that data down the road. You can upload the information to your PC and view it on the screen or use a data browsing program such as a debugger. You might use test equipment such as a multimeter or a logic probe to analyze data signals that are output from your robot's electronics.

Oscilloscope

An oscilloscope is a powerful tool that tests just about everything electronic on your robot, but that power comes at a price. One of these gems in the $400 price range should be able to perform most robot testing tasks. You can purchase oscilloscopes from electronics retail outlets such as Jameco (www.jameco.com), JDR (www.jdr.com), Marlin P. Jones (www.mpja.com), and Mouser (www.mouser.com).

Oscilloscopes quickly take many electronic measurement samples and represent them graphically on a display. Oscilloscopes, such as the one shown in Figure 4-4, are good for checking waveforms, timing sequences, and pulse widths, and for viewing voltage levels over time.

Many digital electronics found on robots output square waves, which are voltages that go instantly on and off. And many sensors output voltage levels that you may need to measure with an oscilloscope to ensure their efficient operation. You need an oscilloscope also for advanced measurements of voltages over a timescale, such as the millisecond-quick on-and-off timing of pulse width modulation; or for verifying sensor timing data, such as data information voltage spikes returned by a sonar sensor to indicate distance.

Other test equipment

Before you attach various devices to your robot, you might just want to make sure they perform as advertised. For example, before using a particular motor, you may want to buy or build a motor-speed control box to find out how well the motor works. With a motor-speed control box, you select a direction and speed. Then you can start the motor out slowly and speed it up gradually. This gives you more control in your testing.

Figure 4-4:
An oscil-
loscope
adds "time"
to your
measure-
ments.

You may also want to experiment by purchasing or making other testing
devices, such as serial servo controllers or a serial I/O box that you can
control from your PC to test various devices such as servos, lights, or
actuators. After you test a device and feel comfortable with the way it
works, you can incorporate it into your robot.

You can purchase motor-speed control boxes and kits as well as serial servo
controllers from any electronics retailer, such as Jameco (www.jameco.com).

Power strips

Power to the robots . . . and test equipment . . . and PC . . . and . . .

Trust me, with all these pieces of electronic equipment in your workshop,
you're going to need lots of power. I recommend getting a surge suppressor
power strip that will protect your expensive electronics and PC. An _uninter-
ruptible power supply_ (UPS) will keep computers and equipment from going
dead in case of a brief power outage.

Also, make sure that you don't overload your electrical wall circuits; you
don't want to burn your house down. And, just to be safe, turn off equipment
when it's not in use.

Machining

If you decide to build a robot from scratch, you won't have nice, prefabricated modules. Instead, you'll have to manufacture many of the parts yourself. Depending on how gung-ho you are, you may need a machine shop full of equipment, such as sheet-metal bending and cutting equipment, milling machines, and lathes.

Most personal robotics work doesn't require a large investment in tools. You may however, want to at least investigate less-expensive miniature machine shop equipment. Check out a desktop milling machine or lathe such as those put out by Sherline Products, Inc. (www.sherline.com).

CAD software

You can't just jump into robot building willy-nilly. You should always take the time to design your robot so you have a roadmap for your journey. In fact, after you've graduated from basic kits, design should be at least 50 percent of the time you devote to building your robot.

Of course, sometimes it's hard to beat a pad of graph paper and a pencil for coming up with a quick design. But when graph paper fails, you might want to look into purchasing inexpensive *computer-aided design* (CAD) software such as AutoCad Lite or TurboCad. You can even find free CAD software, such as Ribbonsoft's Qcad (www.qcad.org), on the Internet. For serious robot designers, some kind of CAD software is a must.

If you don't put the time into designing your robot, you'll find yourself constantly going back to the drawing board, wasting valuable time and money. The first and most important tool you should use when building your robot is your brain.

Testing your circuit board design

When you want to design your own circuit board, using a breadboard is a good place to start. A *breadboard* is a plastic board with holes that are connected vertically by little metal strips. You assemble your circuit by plugging components and wires into the breadboard holes, and then you test the circuit. You can buy breadboards from an electronics supplier such as Radio Shack, Jameco, JDR, or Mouser.

After you have a circuit design that you're happy with, you can construct a perf board or create a custom *printed circuit board* (PCB). A PCB is a fiberglass board with your electronic circuit printed on it in copper. To create your own PCB, you first lay it out with circuit design software, such as that offered by Protel (www.protel.com) or Eagle (www.cadsoft.de). After you're happy with the circuit board layout, you can go to a circuit board manufacturer to get a circuit board custom made.

Integrated development environment

To create software for your robot, you may want to invest in software tools such as an *integrated development environment* (IDE). An IDE is basically a glorified word processor designed for writing code. IDEs allow you to compile your code to machine form and even download your compiled code to your robot.

Sometimes manufacturers provide an IDE free when you purchase a microcontroller development kit. Such a development kit may include a computer board, a cable for downloading, and software for developing code and downloading it to the computer board. The ARobot, for example, uses the Basic Stamp 2 microcontroller (sold separately). The ARobot kit includes a download cable and IDE software to help you write code and download it to the robot.

Although you can use an IDE to write your own code, you can also insert existing code snippets and download them. For that reason, you may want to search the Internet for software function libraries, applications, or code samples so that you don't have to reinvent code or recreate useful algorithms.

A few good places to find IDEs include Microchip Inc. (www.microchip.com), Keil Software (www.keil.com), Parallax Inc. (www.parallaxinc.com), and GCC GNU (http://gcc.gnu.org).

Device programmers

You may also want to invest in a *device programmer,* a box that etches a program or data into a nonvolatile memory chip. You would do this only if you want to make your robot's program permanent and if your computer board allows for such a chip to be plugged into it.

Burning a ROM chip is a more permanent solution than storing data on battery-backed RAM chips. RAM chips can lose program memory if the battery runs out or if there is a brain-scrambling static discharge. Chip devices such as a *programmable IC* (PIC), a *microcontroller unit* (MCU), and an EPROM can often be programmed with an inexpensive device programmer.

Most chips are erased electrically, but some must be erased by exposing them to UV (ultraviolet) light. To do this, you may need an EPROM eraser box, which bathes the EPROM in a strong, short wavelength light for a few minutes to erase the program memory.

The short wavelength light used to erase program memory is very harmful to your eyes, so don't look inside an EPROM eraser box during this process.

You need a device programmer only if you plan to burn chips. In Chapter 10, you find out how to download a program from a PC to your robot for the ARobot kit, which doesn't need a device programmer. You will rarely if ever need to burn chips with a device programmer because most robot brains are advanced enough to have nonvolatile memory onboard, where you can write and store a program. You need a device programmer only if you plan to create your own special electronic circuits from scratch.

How to Solder

Chances are you'll need to know how to solder sooner or later in your robot-building career. *Soldering* (perversely pronounced "soddering") involves a material called *solder* which melts when placed on a hot object; the melted solder cools and forms a bond between two items.

First, you'll need a soldering iron with a soldering station. A *soldering station* holds your hot soldering iron and keeps your solder and tip cleaner organized. Purchase a small 15- to 30-watt soldering iron made for electronics and a soldering station. Also buy thin .032-inch-diameter rosin-core solder. You can purchase these at your local Radio Shack and other places.

Don't use a big soldering iron and the big ¼-inch, acid-core solder used for plumbing, which are typically found at home improvement stores. If you do, you might damage sensitive electronic components. Use rosin-core solder in your robot projects.

Figure 4-5 shows the basic process of soldering. Figure 4-6 zooms in on the process.

Soldering 101

The best technique for soldering is simple, so repeat after me: Heat the metal, not the solder. For example, you heat the metal of a component pin and the metal of a circuit board pad simultaneously, and then you touch the tip of the rosin-core solder to the pad or the pin, but not to the iron. If you have sufficiently heated the two metals (the pad and the pin), they will heat the solder, which then flows quickly to both the pad and the component pin. See Figure 4-7 for an example of good and bad solder joints.

It's also important to know which piece to solder to which other pieces. For example, a *pad* is the little copper metallic doughnut around a circuit board hole that you can put a component pin through. A *trace* is one of the copper lines on the circuit board. You usually solder a component to a pad, not directly to a trace.

Undoing solder mistakes

If you do make a mistake with solder, you'll be glad to know that you can undo a bad solder. One method is to just heat up the bad solder and then suck it away with the solder sucker.

Another way to remove unwanted solder is to use *copper braiding*. You put the braiding on top of the solder that you want to remove and heat it with your soldering iron. The copper braiding absorbs the unwanted solder. You then discard the used copper braiding.

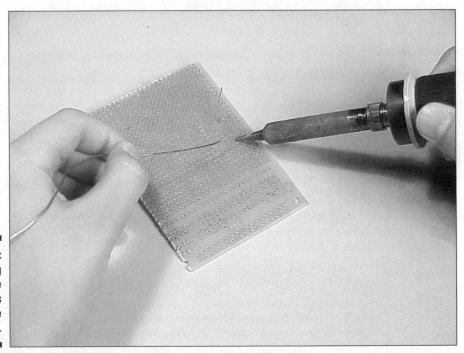

Figure 4-5:
Soldering requires the right tools and a little skill.

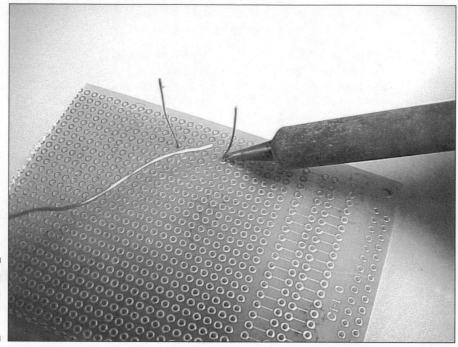

Figure 4-6:
Here's
soldering,
close up.

Soldering tips

Because soldering is an important robot-building skill, you'll want to master the basic techniques quickly. Here, then, are some essential tips to good soldering:

✔ Remember the old joke about knowing which end of the soldering iron to hold on to? Seriously, a soldering iron can burn you or cause a fire. Liquid solder can cause severe burns too, so always use caution when melting solder.

✔ When you solder something, it will remain hot for many minutes. Always grab parts with pliers to avoid getting burned even after the soldering iron is removed.

✔ Purchase the correct solder type and width, as well as the correct soldering iron and tip. Think small tip and thin solder.

✔ Some soldering kits include training materials to help you master the art of soldering. Although people can tell you how to solder, good soldering requires hands-on experience. Take the time to solder a few cheap test components into a test prototype board to get your technique down before using your skills on somewhat more costly robot parts.

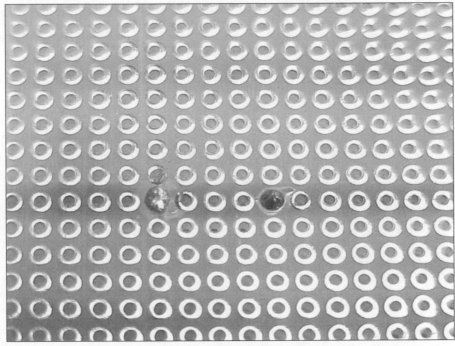

Figure 4-7:
A bad solder joint (on the left) — and a good one (on the right).

✔ If your solder looks like a clump of wadded-up aluminum foil, you've soldered it incorrectly. The solder should look smooth and shiny and must cling to both items (for example the component pin or wire and PCB pad) to make a good connection.

✔ Incorrect soldering (such as cold solder joints) can lead to all sorts of problems that can be hard to track down.

✔ Be careful not to apply your soldering iron for long periods of time. Otherwise, you can damage sensitive components or burn up a circuit board trace. You should solder quickly so that your components or trace don't stay hot for too long.

✔ You should always make a mechanical connection before making a solder connection. For example, check to make sure the component pin actually touches the wall of the pad hole before you solder it. This will ensure that your soldering goes quickly and smoothly and will help to keep a solder joint from "bridging" to the pin and separating.

✔ You may want to flux before soldering to get a cleaner solder. *Flux* is a pasty, greasy, oily substance that helps to clean the metallic surfaces being soldered. It also helps you produce smooth solder joints that adhere well to pin and pad surfaces.

Rosin core solder used with robots has flux conveniently within the solder, so fluxing is usually not necessary. However, for a dirty or older solder joint, where the flux may have dissipated, you may want to brush a little flux on to help you rework the old solder joints and make them clean and smooth again. You can purchase a small can of flux at just about any electronics store.

✔ Only experience will tell you if you have soldered correctly, so ask an experienced soldering friend to check your work. Doing so can save you hours of debugging time later.

How do you keep it so clean?

As somebody probably said, take care of your tools and they'll take care of you. You should perform preventative maintenance and regular cleaning to prolong your soldering iron's life. Don't let your soldering iron tip get dirty. While your soldering iron is hot, clean the tip often with a bit of tip cleaner and a moist sponge or paper towel. Remember: A dirty soldering iron will make terrible solder joints.

While your soldering iron is hot, you may want to *tin* the tip with solder to get it shiny and clean and to remove any dross or rosin. Tinning also helps prevent oxidation. To tin the tip, get your soldering iron hot and then coat the tip with solder. Your tip should look like chrome or silver.

Always unplug your soldering iron when you're finished using it to help prevent oxidizing and burning up the tip.

Anti-Static Considerations: Can't Touch This!

Have you ever walked across the carpet on a humid day and then touched a doorknob, only to get a painful zap? What you experienced is *static electricity,* which is created when you build up voltage potential on your body and it discharges to a grounded object.

Now pay attention: Static electricity is dangerous to your robot's electronics. Here's how to avoid it.

Mr. Owl, how many volts does it take . . . ?

When static electricity discharges, it may involve hundreds or thousands of volts or just a few volts. Many electronic components used in robotics have a tolerance range of only 6 volts or less. Any discharge more than 6 volts may destroy your precious electronics. And even if your electronics aren't destroyed by a discharge, they may be seriously damaged, causing malfunctions or a drastically shortened chip life.

Now electricity, as Ben Franklin found out, can be deceiving. It may take several hundred volts for a person to feel such a discharge, and it may be several thousand volts before you'll ever see a spark fly. In other words, static discharge from your body can be a silent and invisible robot killer.

The honorable discharge

Some materials that tend to build up static electricity are sweaters, paper, hair, wood . . . well actually, just about everything. So how do you avoid static electricity? A good procedure to follow is to discharge yourself by touching some bare metal, such as a desk leg or a grounded metal plate, before touching anything electronic. Also, when you hand an electronic part to someone else, it's a good practice to touch that person's hand first to discharge any electricity. (Several robot builder romances have begun this way).

Don't touch metal robot parts without discharging yourself and always handle PC boards by their edges. Although touching the frame of the robot is usually a safe way to handle it, don't touch anything metallic that may be connected to the robot's electronic circuit; otherwise, a discharge may cause damage to the electronics.

How cautious should you be?

If robot-handling guidelines make you cringe, you can decide not to be so OSHA-regulation paranoid and just be careful where you touch. For example, touching the ARobot's frame is okay, but touching a finger on the exposed Basic Stamp 2 brain or other circuit parts is bad.

The precautions you take depend on how foolproof you want to make your robot. If you know that you'll be the only one who's going to touch your robot, and you know where to touch and where not to touch, you and your brain are your robot's protection. But if you worry that others are likely to touch where they shouldn't, you should put up barriers around those sensitive areas.

There are an infinite number of considerations in protecting your robot from static shock, and it takes only one bad discharge to kill your robot. So decide how you want to protect your robot against static discharge and determine whether you will go through the trouble to put up barriers around sensitive areas or just have a gentlemen's agreement not to touch sensitive areas.

As you design your robot, you need to include a plan for handling the robot without destroying it with static discharge. That means you may need to shroud your robot's electronics with a "skin" to prevent accidental touching and discharges. This skin could be a simple plastic enclosure or an elaborate metal contraption.

You can purchase a can of static-resistant spray and spray it on your desk and carpet to minimize static electricity. You can also wear a grounding strap and use an antistatic mat while handling electronics.

Finally, you may want to encase your robot's electronics inside a plastic project box mounted on your robot to avoid any inadvertent touching that could accidentally damage the electronics with static electricity. Project boxes are easy to find at electronic shops and mail-order suppliers of electronic components.

A Robot Library

All robot builders must have certain information tools. That's why you should always leave room in your workshop for a little robot library. You can even program your robot to dust the books off periodically.

Having information at arm's length that you can locate quickly can be a lifesaver. In fact, keep this book close at hand, read it often, and recommend it to your friends. The author will thank you.

You should buy a few useful books right off the bat to begin your robotics library. Here are a few of my favorites:

- ✔ *Mobile Robots: Inspiration to Implementation* by Joseph L. Jones, Anita M. Flynn, Bruce A. Seiger (published by A. K. Peters Ltd.)
- ✔ *Build Your Own Robot!* by Karl Lunt (published by A. K. Peters Ltd .)
- ✔ *Robot Builder's Bonanza* by Gordon McComb (published by McGraw-Hill/TAB Electronics)
- ✔ *Getting Started in Electronics* by Forrest Mims III (published by Radio Shack Corp.)

As you work on ever more sophisticated robots, manuals and software language tutorials will help you expand your knowledge.

Visit a discount or used bookstore in your area to look for books you might need. The Internet is also a good, inexpensive source for information. You might also try the library or borrow books from a friend.

Magazines have information about current robotic trends and new robot kits that's usually more up to date than what you find in books. Check out any of the following: *Electronics Now, Circuit Cellar Ink,* or *Nuts & Volts* magazine.

Robot sources on the Web

The robotics industry is always changing, and what is cool and new today may be obsolete tomorrow.

Scan a few Web sites such as Robots.net (www.robots.net) to keep up to date on the latest in robotics news.

You might also check out robotics stores such as The Robot Store (www.robotstore.com) or electronics stores such as Jameco (www.jameco.com). Checking out some of these sources might help you find a new gadget or robot kit or give you an idea that may save you time and effort.

Chapter 5

The Nuts and Bolts of Robot Building

*J*ust as it's a wise cook who knows about ingredients, it's a wise robot builder who understands about the parts that go into a robot before starting to build. In this chapter, you get the cook's tour of all the great stuff you'll be working with when you enter the world of programmable robot building.

Pieces and Parts

In a way, you can think of robot parts in terms of human parts. All robots have the following:

- **Senses:** Devices that detect the world around your robot, such as whisker sensors, cameras, and water, voltage, or smoke detectors.

- **A brain:** A device that allows your robot to react or make decisions. Examples include a CPU and a bump switch.

- **A nervous system:** Devices or parts that enable communication or control various other parts, such as data cables and connectors, power cables and connectors, and hydraulic lines.

✔ **Mechanisms for movement and locomotion:** On a robot, these include motor-driven wheels, motor-driven legs, servo-driven grippers, and air muscles.

✔ **Outputs:** Like a mouth on a human, these are devices that give feedback, such as lights, an LCD monitor, or a speaker.

✔ **A power source:** Energy that feeds and keeps your robot going, such as batteries, fuel for driving a fuel cell or generator, or even compressed air.

✔ **A body:** The construction that ties your robot together, that is, the frame or chassis.

What's It All Made Of?

Just like you, every robot has some kind of body. In the case of robots, the body is called a *chassis*. A robot's chassis has to be made out of some kind of material, and some materials are better than others, depending on how you want to use your robot. Wooden robots do not make good undersea rovers, for instance.

In this section I discuss some of the common materials used in robotics and explain when it may be better to use one material over another.

Heavy metal

If the three little pigs were building a robot, the one who wanted the strongest material would build with metal. Many robot builders prize aluminum for building their robot's structure.

Common construction methods for metal pieces involve cutting and then bolting or riveting pieces together. In addition, you sometimes have to bend metal to match your robot's design. For that reason, you need a metal that can be cut, bent, and drilled easily. Sheet aluminum is light and strong and easy to work with. Other metals, such as steel, are too heavy, too hard to bend, and difficult to cut or drill.

Metal is a good material to use if you plan to ground your chassis. You can ground your robot's chassis in the same way that you ground a car chassis — by attaching the negative side of the battery to the frame. Frame grounding your electronics can help protect your robot against static electrical discharge. But you should still use standoffs for your electronics so that your circuits don't short against the metal and so that only the negative side is connected to the frame. (*Standoffs* are like little risers that isolate, or lift up, the bottom of your printed circuit board.)

It's not a good idea to use bare metal around water because it may rust. You can avoid rust by painting metal parts. Also, water can short out a battery.

Play it safe when working with metal because it has sharp edges that can cause serious cuts. You may want to file or sand sharp edges or cover edges with grommets or rubber edging.

However, just as building a house of brick costs more than building with straw, metal can be more expensive than a material such as wood.

One word — plastic

Plastic can be better than metal for some purposes. Plastic can be strong, though usually not as strong as metal. But plastic has the benefit of being lighter than metal, and lighter-weight robots may use less power to move around. Plastic is usually water and rust resistant and therefore may be good for building robots that come in contact with water or moisture.

You can purchase plastic in sheets such as PVCs (polyvinyl chloride, a kind of plastic often molded into pipes and used in plumbing applications) that may be cut, heated, and molded. Plastic can be milled or cut from a block. Kit robots are usually made in plastic because it can be cheaper and easier to design and mold in plastic when manufacturing large quantities.

Construction techniques with plastic include gluing, drilling, and bolting.

It was good enough for Pinocchio . . .

Few robot builders use wood for their robot chassis, but don't discount it entirely. Here's why:

- ✔ Wood is easy to cut, drill, and put screws into.
- ✔ Wood is usually the cheapest construction material.
- ✔ Wood doesn't conduct electricity, so there's little hazard of shorting out your electronics.

You can even find some professionally made kits that call for making a robot out of wood. Plywood or shelf boards are common wood construction materials. Construction techniques include gluing, bolting, and using metal L brackets to connect one piece of wood to another.

Nuts and Bolts

You can form attachments in many ways and so can your robot. You can use glue, brackets, or nuts and bolts, for example, to attach things such as motors to your robot's chassis.

Of these techniques, using nuts and bolts is probably the most secure way to attach one thing to another on your robot. However, you should be aware of some facts about them.

Double standards

Nuts and bolts come in many sizes and lengths. It doesn't matter whether you choose metric or standard (English) measurements. What does matter is sticking with either metric or standard, if possible. Otherwise, you'll need twice the number of tools to work on your robot and you'll always be trying to figure out which type of tool you need.

You should use standoffs when attaching electronics with nuts and bolts. Always match the thread size of the nut, bolt, and standoff. For example, you might decide to use a standard #10 for nuts, bolts, and standoff sizes on a project to simplify things.

Don't fall apart on me

There are good vibrations and bad vibrations. Good vibrations make you feel happy, and bad vibrations can cause robot problems. Robotic platforms usually vibrate a lot, which causes nuts and bolts to become loose. To avoid this problem, you may want to use lock washers. Another technique is to use Loc-Tite adhesive to secure your bolts. Loc-Tite comes in different strengths: permanent, semipermanent, and a few variations in between. Look for the kind of Loc-Tite that will hold securely, but still allow you to remove the nut or bolt later if you want to disassemble the robot and reuse the parts.

Avoid the permanent type of thread-locking liquid if you think you might want to make changes to your robot some day.

Be sure to follow the Loc-Tite instructions carefully so that you get the degree of permanence you need. Also follow safety instructions or you may glue your finger to your robot, and you don't want to go there.

Motors for Locomotion

Your car needs an engine to run. Even your can opener probably has a little motor inside to spin the can around. Your robot also needs something to make it run.

An *actuator* activates a mechanical device and is basically anything that causes movement on your robot. Motors, such as the motors that drive a robot, are the most common type of actuator. Motors come in several varieties such as the gear motor and the servo motor, which you'll hear more about in this section.

Motors vary in power, speed, accuracy, and power consumption. Some motors have shafts that rotate continuously; other types turn less than a complete rotation.

Motors can add a lot of weight to a robot. Lightweight aluminum motors or even plastic motors might save your robot a lot of power in the long run.

Here, then, is everything you ever wanted to know about motors (and more).

The useful DC gear motor

The two types of motors that you're likely to use in your robotic adventure are DC gear motors and RC servo motors. The most common motor for robotics is the DC gear motor, which works by gearing down a fast DC motor to make the motor turn at a slower speed and give the motor a higher torque suitable for robot locomotion. The Soccer Jr. robot uses a DC motor along with separate gears to effectively create a gear motor.

How much motor do you need?

Motors use power, so the fewer motors you have on a robot, the better your robot's power efficiency.

That said, robot builders sometimes design their robot to have two drive motors for locomotion, which makes the thing move around like a high-powered tank. Other builders design their robots to use one motor to drive the robot and another one for steering.

Any design with one drive motor will result in longer battery life. The two-drive-motor design, however, may provide other benefits at the cost of battery efficiency, such as a tighter turning radius.

A DC gear motor is basically a regular DC motor with a special gear box attached to the output shaft, as shown in Figure 5-1. Your robot's electrical drive circuitry can control the DC gear motor to rotate the wheels of your robot for locomotion.

You can get a DC motor without a gear head, but generally these are too fast (around 15,000 RPMs is typical). For a robot to move at a reasonable rate, you have to gear down a DC motor to about 30 to 80 RPMs. When you gear down a DC motor, you get a slower speed and plenty of torque.

ARobot, which you work with in Part III, uses a DC gear motor for locomotion.

Controlling your motor

To get your robot to go fast or slow, you need to control the speed of its motor. To regulate the motor, you vary its on time and off time, also known as the *duty cycle,* to simulate changes in voltage level, as shown in Figure 5-2. This method of varying the duty cycle to regulate motor speed is also known as *pulse width modulation* (PWM).

A duty cycle of 50 Hz or faster (20-ms pulse time) will work for the motor, but to get above human hearing so that you won't hear a hum, you must increase the duty cycle to over 20 KHz. This means the full pulse time must be 50μs or faster.

Figure 5-1:
A gear
motor.

As shown in Figure 5-2, the percent simulated voltage roughly equates to the percent speed of the motor. So a 50% duty cycle allows the motor to rotate at approximately half speed. You can modify the pulse width On time to accommodate other duty cycles.

PWM is the preferred way to vary motor speed because it simulates voltage drop but does not increase the current as a resistor would. (To find out more about resistors, see the "Electronics Primer" section later in the chapter.) Because PWM doesn't increase the current, your motor's coils are safe from damage.

PWM has more to do with torque control than with exact speed control. That means the speed control may not be exact. For example, if you give your motor a 50% duty cycle, your robot may actually go at only 40% speed. You may have to experiment to verify the speeds you get at various duty cycles. As a car dealer would say, your mileage may vary.

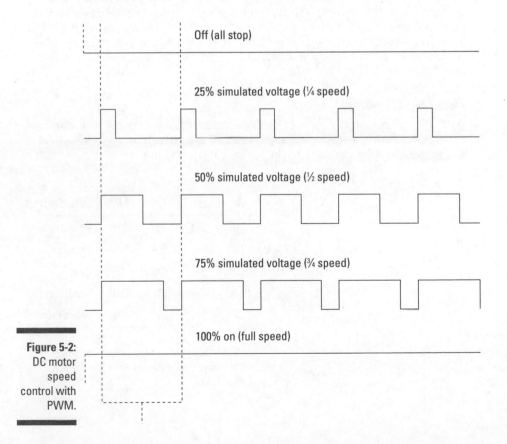

Off (all stop)

25% simulated voltage (¼ speed)

50% simulated voltage (½ speed)

75% simulated voltage (¾ speed)

100% on (full speed)

Figure 5-2:
DC motor
speed
control with
PWM.

To use PWM, you need a circuit. One of the most popular motor controller circuits is an H-bridge circuit, shown in Figure 5-3. An H-bridge circuit turns a motor on and off, allows a computer to control a motor's direction (forward or reverse) and regulate speed, and may even provide a breaking mechanism. A DC gear motor's rotation direction is usually controlled with an H-bridge circuit.

A computer can't control a motor directly for a several reasons. First, a computer doesn't output enough power to drive a motor. Second, a computer can't control direction because it has only outputs. Third, motors are noisy, electrically speaking, and would quickly damage a computer. Essentially, the computer sends signals to the H-bridge to tell it to go forward, reverse, brake, or add speed. The H-bridge then steps up the voltage and power for the motor. Figure 5-3 shows how a motor's direction is controlled based on electrical signals from the computer.The H-bridge circuit also isolates the computer from destructive voltage spikes and noise, which arise mainly from motors. In addition to using an H-bridge circuit, you might want to have two sets of batteries: one for your electronics and another for your motors. One way to do this is to have one 12-volt battery and step down the voltage with a 7805 5-volt regulator chip. This is inefficient but simple. A more efficient method is to have a 12-volt battery source and a 5-volt battery source to avoid the power loss in voltage regulation.

How far will you go?

With a DC gear motor, your robot will go far. Just how far takes a bit of measuring. Measuring distance traveled is easy if you attach an encoder wheel and some counting circuitry to the motor's output shaft.

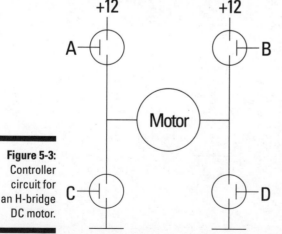

A	B	C	D	Function
ON	OFF	OFF	ON	Forward
OFF	ON	ON	OFF	Reverse
ON	ON	OFF	OFF	Break
ON	OFF	ON	OFF	Short circuit*
OFF	ON	OFF	ON	Short circuit*

*Warning: This configuration causes a short and could damage the circuit and motor.

Figure 5-3:
Controller circuit for an H-bridge DC motor.

The calculated, gory details

Put your math thinking cap on for this one. To measure distance traveled with a chopper wheel, you have to make some calculations. For example, suppose that a chopper wheel is attached to the DC motor side, which is turning at 15,000 RPMs. Also suppose that the chopper wheel has 100 slots, your DC motor has a gear head with 1:1000 reduction, and the wheel connected to that gear head motor is 6 inches in diameter.

You first need to calculate the amount of rotation traveled: 15,000 pulses / 100 per revolution = 1500 revolutions. 1500 revolutions / 1000 for gear reduction = 1.5 revolutions of the output

shaft. The wheel is 6 inches in diameter, so 1.5 revolutions means a distance traveled of roughly 6 inches diameter x 3.14159 x 1.5 revolutions = 28.27 inches traveled. Therefore, the 28.27 inches / 15,000 pulses read means that each pulse equals .0001884 inch.

When you design your robot's chopper wheel, you'll have to take these measurements into account. To make things simpler, you could move the chopper wheel and circuitry to a wheel with 19 slots, so that each pulse measures approximately 1 inch (6 x 3.14159 = 18.84, and a 6-inch wheel is almost 19 inches in circumference).

The two basic types of encoder wheels are directional and chopper. *Directional wheels* look like round black-and-white checkerboards and are typically used to take compass readings. *Chopper wheels* coupled with an infrared sensor can help you count distance traveled. The number of holes in a chopper wheel determines the length of something called a *pulse*. By giving your computer information on the number of pulses that the chopper wheel has moved, the computer can count and then calculate the distance traveled.

ARobot has a chopper wheel on the drive motor to help measure the distance traveled.

The versatile RC servo

Robots are supposed to be autonomous, so what's all this about remote-control servos? Does it mean your robot is now being controlled remotely rather than autonomously? No. RC servos are the actuator boxes used in remote-controlled cars and airplanes, but they work well for robotics as well. These servos are not really remotely-controlled as their RC servo name implies. Instead, they're just servos produced by the RC industry that you can use for your robotics.

ARobot, which you'll work with in Part III, uses an RC servo for steering.

Inside an RC servo is a DC gear motor with a tiny computer that allows it to accept positioning commands and to point the head of the motor in various directions. RC servos usually have a limited range of motion of only 120 degrees or so.

You can use RC servos like the one shown in Figure 5-4 to point sensors, pan and tilt a camera, move leg and arm joints, turn a head, and open and close a gripper.

RC servos usually run on about 4.5 to 6 volts and are power efficient, typically drawing only around ¼ amp. RC servos also have lots of torque. You can buy servos ranging in size from micro servos (when size is a consideration) up to big high-torque versions. You can also find servos with more range of motion, such as 180 degrees.

RC servo motors are controlled through a repeating *pulse width positioning* (PWM) signal, as shown in Figure 5-5. Each servo has its own timing considerations, but typically a 1-ms to 2-ms pulse repeated every 25 ms or so will tell the servo where to position its head. For example, if you send a 1-ms pulse repeatedly, the servo positions itself all the way to the left. If you send a 2-ms pulse repeatedly, the servo positions itself all the way to the right. A 1.5-ms pulse positions the servo to the middle.

By choosing a pulse width somewhere between 1 ms and 2 ms, you may choose other servo positions. In addition, some servos may have a wider range of motion, so that 0.5 ms might move the servo farther to the left and 2.5 ms, farther to the right.

Check with your servo's specifications to make sure it allows extra range of motion before you try this extended range. Otherwise, you could destroy your servo.

Figure 5-4:
An RC
servo.

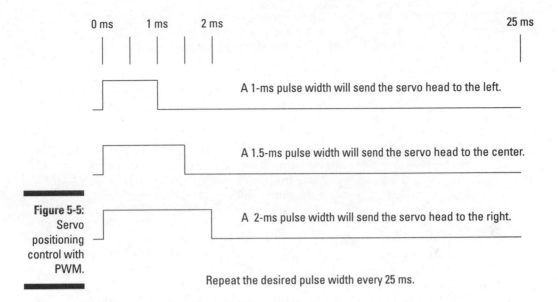

0 ms   1 ms   2 ms   25 ms

A 1-ms pulse width will send the servo head to the left.

A 1.5-ms pulse width will send the servo head to the center.

Figure 5-5:
Servo
positioning
control with
PWM.

A 2-ms pulse width will send the servo head to the right.

Repeat the desired pulse width every 25 ms.

Some RC servos can be hacked to enable the servo head to rotate freely 360 degrees. A rotating RC servo basically makes your servo a DC gear motor with computerized direction and perhaps even speed control built in.

To hack a servo, you remove any limitation "stops" that keep the head from turning freely and create a resistor network with the feedback circuit. For more information about hacking servos, research the topic on the Internet. Not all servos can be hacked, so you might want to also investigate which servos can be hacked before you buy one.

Many of today's smaller robots use hacked RC servo motors as the drive train. Some robotic kits even use hacked servos for locomotion. A hacked servo turned into a DC gear motor is usually a less expensive, more compact, and more controllable solution than buying a regular gear motor.

The power-hungry stepper

A stepper motor is a unique motor. Unlike a DC motor, which you just turn on and it begins turning, a stepper motor has various wires that must be electrically pulsed sequentially with a computer or an electronic circuit to cause the motor to turn a few degrees, one step at a time. Steppers are usually used in a robot to position a sensor or control rarely used grippers — they're not used as the robot's drive train.

I recommend avoiding stepper motors for robot locomotion because they're power hungry and usually have little torque for the power they consume.

Other actuators for locomotion

If you want to operate your robot without a power cord attached, power efficiency is important and you'll need to avoid using actuators that drain a lot of power. If you use a power cord (called a *tether* in robot speak), however, power efficiency is less crucial and you might consider using some alternate types of actuators:

✔ *Nitinol wire,* also called *memory wire* or *muscle wire,* shrinks as heat is applied and relaxes as it cools. Because the wire is resistive, you can apply power to make it hot and therefore contract or remove power to let it cool and therefore expand. It takes a second or two to expand and contract depending on how much or little heat or cold is applied. With progressive cooling, you may get quicker cycle times, but for the most part Nitinol wire can be slow to respond. Also, because Nitinol wire acts as a resistor, and resistors waste power, it is very power hungry. Nitinol, or NiTi, wire is like a muscle: instead of rotating like a motor, it pulls. Nitinol wire doesn't have much torque, but you can use more than one wire to increase its pull. You will almost never see Nitinol wire in a personal robot because it just can't move much weight and the cycle time is slow. If your robot is tethered, though, you may consider using Nitinol wire.

✔ Robotics does not always use only electronic actuators. Air muscles are another option. *Air muscles* consist of a rubber bladder of air encased in a braided nylon tube. When you add air, the bladder expands, causing the ends to draw in. Releasing the air causes the muscles to deflate and relax back to their full length. Air muscles provide a lot of power, but they also require a lot of work. You must add electronic air valves and provide for constant air pressure, as well as attach special air pressure tubes and connectors. For many people, the added complexity, weight, and expense may not justify using air muscles for a personal robot. In addition, air muscles may not last as long as other actuators. However, air muscles can lift a lot of weight.

✔ If you know your Greek, you'll know that *hydra* means water. Therefore, *hydraulics* relates to something that's operated by the force of liquid (think Hoover Dam . . . think hydraulic brakes). With hydraulics, you can lift just about anything, but the cost and weight of such a setup means hydraulics are used mostly for huge industrial robotics.

Making Sense of Sensors

Sensors are the devices that help your robot detect the world around it: They provide the equivalent of touch, smell, sight, and hearing to your mechanical friend. A sensor works by detecting a change or a measurement in the world around it and reporting that change or measurement back to the computer's brain.

For sensors to be of any use to the software that runs your robot, they must be attached to the robot's brain — often in a particular place and direction.

For example, a camera must be aimed in the direction in which you want it to shoot pictures.

Some sensors, such as an electronic compass, must be level to work accurately. Other sensors, such as moisture detectors, should be placed low or at the level where ground water can be detected. Light sensors should probably be placed on top, where light can be sensed easily.

I go into more detail about the different kinds of sensors and their applications in Part IV, where you get to find out about various sensors, build circuits with them, and write software to use them.

Electronics Primer

Just about every robot has electronics that control its actions. Just as to drive a car you hardly need a degree in mechanical engineering, you can build simple robots without knowing that much about electronics. Many robot kits provide building blocks and prebuilt electronics that you just slot into place. For those who want to become savvy robot builders (and to prevent a brownout on the Eastern seaboard), I cover electronics basics here.

Resistors resist

Resistors, such as the one shown in Figure 5-6, look like tiny hot dogs with stripes. *Resistors* resist current flow and are used as current-limiting devices that help keep sensitive components from being destroyed by bursts of high current. For example, you may want to add a 330-ohm resistor inline with an LED to limit electrical current, which could destroy the LED.

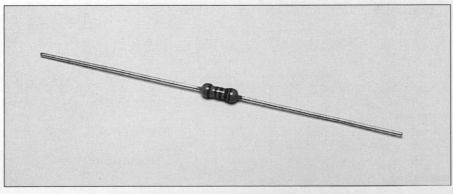

Figure 5-6:
A resistor
is one of
the most
common
electronic
components.

Essentially, a resistor converts resisted electricity into heat, so you may notice that a resistor component gets warm or even hot when electricity is applied. For that reason, resistors make great heaters. In fact, many electrical heaters are just large-scale resistors.

Be sure not to allow anything flammable to touch a resistor when power is applied. Otherwise, the resistor may catch fire.

Another type of resistor used with robotics is a variable resistor called a *potentiometer,* or *pot.* A pot is like a volume control knob. You can dial up the resistance you want.

When you assemble your electronics, and you have to solder in a resistor, you should be sure you're using the right value of resistor. So how do you figure out the value of the resistor? To read a resistor's resistance value, you can use a multimeter or read the stripes painted on the outside.

The stripes indicate the resistance value and tolerance of the device, but you have to know how to make sense of them. Figure 5-7 lists the colors you find on resistor stripes, along with their values, which are ordered from 0 to 9.

Because it's difficult to remember values of the colors of the stripes on resistors, every good electronics engineer uses a BBROYGBVGW mnemonic. Here's one: Black Birds Ruin Our Yellow Grain, Butchering Very Good Wheat.

Every resistor has three or four stripes. The first two stripes indicate the first two significant digits of the resistance value. The third stripe is the multiplier stripe, where black equals 10^0 (which is 1), brown equals 10^1 (which is 10), red equals 10^2 (which is 100), and so on. The fourth stripe indicates the tolerance: gold is 5%, silver is 10%, and no fourth stripe is 20%.

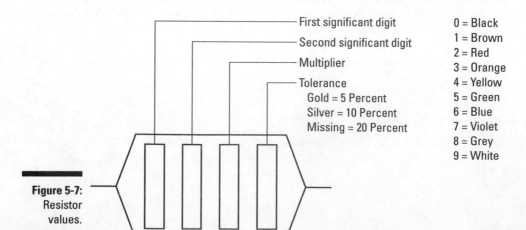

First significant digit
Second significant digit
Multiplier
Tolerance
Gold = 5 Percent
Silver = 10 Percent
Missing = 20 Percent

0 = Black
1 = Brown
2 = Red
3 = Orange
4 = Yellow
5 = Green
6 = Blue
7 = Violet
8 = Grey
9 = White

Figure 5-7: Resistor values.

Motors and resistors don't mix

It's generally not a good idea to control motor speed by using a resistor or a potentiometer.

A resistor will cause the coils in a motor or solenoid to get overheated, so putting a resistor or potentiometer inline with a motor is not a good idea. Also, using resistors on robots is like throwing power away because resistors throw off power in the form of heat.

Instead of using a resistor to change motor speed, use a fast on-and-off variable duty cycle circuit to simulate voltage level drop. This will cause the simulated voltage to drop without increasing the current.

For example, a 330-ohm resistor usually used to current-limit a 5-volt LED is orange, orange, brown, gold. The first orange stripe is 3, the second orange stripe is 3, the third stripe, brown, is times 10 (or 33 x 10 = 330), and the fourth stripe, gold, is 5% tolerance, or 330 +- 33 tolerance. Therefore, the 330-ohm resistor's actual value could be some value between 297 and 363.

Resistors come in many different packages — some small, some large and bulky. The size mainly has to do with the amount of heat they dissipate. Resistors also come grouped together as *Single In-Line Packages* (SIPs), where many resistors (typically up to nine) share a single package.

Take charge with capacitors

Capacitors, such as those shown in Figure 5-8, build up an electrical charge when voltage is applied. You can think of them as being like tiny rechargeable batteries.

Capacitors, or caps, are useful for a variety of purposes. Small ceramic capacitors such as the 0.1uf are good for filtering voltage spikes; you'll find them near sensitive memory or CPU chips. Even smaller ceramic or mica caps, around 25pf, are useful as a support component for CPUs to help keep them ticking. Larger capacitors such as 10uf up to 6f capacitors are good for smoothing and leveling voltage or sustaining power for a time after external power is removed. You usually find larger capacitors in power-supply applications.

Because capacitors store electrical charges, you should take care when handling them to reduce the chance of electric shock.

So how do you figure out the value of the capacitor? The process is similar to figuring out the value of a resistor.

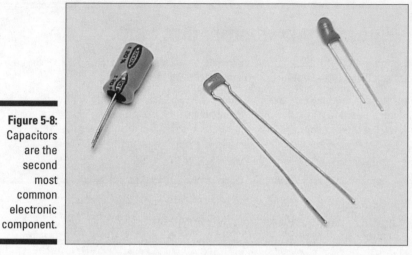

Ceramic disk capacitors have a three-digit number. The first two digits represent single numbers and the third digit is the multiplier. If a letter is present, it specifies tolerance. The letter *K*, for example, means 10% tolerance (the value could be plus or minus 10% of the value marked). The value is stated in picoFarads. For example, the number 104K, shown in Figure 5-9, means 10 with a multiplier of 4 — which is 10^4, which is 10,000 picoFarads, or 1.0 microFarads — +- 10% tolerance.

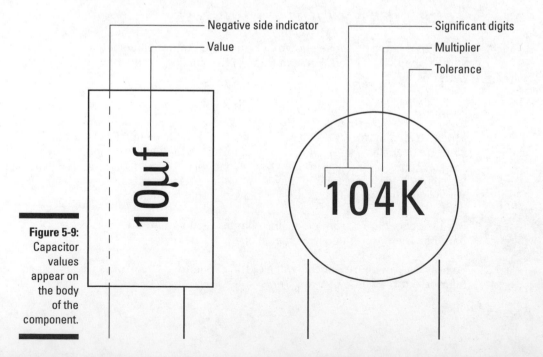

Think modular

Being able to swap out parts on a robot is essential if you want to upgrade parts or replace broken parts. To prepare for swapping out parts, you should design your robot in a modular way. Instead of hard-soldering wires from one part to another, use connectors and cables. This allows you to easily connect and disconnect parts.

Using connectors and cables may add to the cost of your robot, but it adds also to the convenience of making modifications and repairs.

If a capacitor is polarized (designed to be used in only one way), a minus or plus sign near one of the leads signifies which is which. Most large capacitors are polarized; most small ones are not.

Never use a polarized capacitor that has been used backwards. It may burst open or worse.

Cable, connectors, and wires

You might use cables and connectors for a variety of tasks on your robot, such as providing power or connecting the robot to your PC to download programs. Lots of cables are available, but you should try to use standard solutions if possible.

Data cables use data connectors, and power cables use power connectors. Data cables usually have ten or 25 thin pins and are gold plated to reduce data loss. Power connectors are big and bulky and usually have just two big fat pins. Data cables have very low amperage, so they're thin, whereas power cables carry lots of amperage and are therefore thicker.

A good choice for data cables are standard, two-row, header-pin connectors and ribbon cables. For power cables, use 6 to 20 gauge wire with Molex connectors, depending on the amperage that the wire will carry. Many battery packs use flat-tab connectors or 9-volt connectors, so design your power cables accordingly. Always try to use the correct type of connectors and the appropriate type of cables, based on your robot's design, between electronics parts to allow for modular assembly and disassembly.

You choose the wire's thickness, or gauge, based on how many amps you plan to use. Wires that you will bend often should be stranded wire rather than solid wire. (Solid wire breaks if bent too many times.) To prevent brownout, wire that will carry a large amount of power should be of a thick enough gauge to handle many times the power that you expect it should need.

To learn more about wire gauges and current ratings, visit the following Web site:www.powerstream.com/Wire_Size.htm

Be careful: Using the wrong gauge wire could result in wire meltdown.

Semiconductors

A *semiconductor* is a silicon device that can change or control the conductance of electricity by various means such as amplification or switching. In your everyday life, you encounter semiconductors in clocks, radios, phones, and televisions. A semiconductor component may have only one function or may be able to perform many functions.

I don't know if you've noticed, but a plethora of semiconductors are out there. Maybe even two plethoras. I won't even try to explain all the various semiconductors available or everything each semiconductor can do. A good source for semiconductor spec sheets is Jameco's Web site (www.jameco.com).

What follows should give you enough understanding of the world of semiconductors to begin to dabble in robot building.

Diodes

Diodes are semiconductors that control the flow of current so that it goes in only one direction. *Light-emitting diodes* (LEDs) are commonly used in robotics as feedback that power is on or as in indicator that the robot is processing something.

Because an LED allows current to flow in only one direction, you must insert the device into a circuit in the correct orientation for it to function properly. A diode usually has a strip on the anode (negative) end to help you position it correctly.

Transistors

Transistors are usually small, three-pin devices that work like a switch or something called a linear amplifier. Basically, when you apply current to the base pin of the transistor, the current conducts, or amplifies, in a linear fashion across the other two pins.

Using the transistor as a switch is similar to using it as a linear amplifier, except you're more interested in the threshold at which the transistor is conducting rather than how much it's amplifying. Ideally, such a transistor would have a less gradual range from off to on so that it switches between the two states quickly.

Pinouts

A typical cable has several wires inside, with each wire connected to pins on the connectors. A cable can have different connectors on each end, and the wires can be connected to the pins in any fashion. The definition of this internal cable wiring is referred to as the cable's *pinout.* A pinout diagram describes each wire and indicates which pin on each connector that the wire is connected to.

In a transistor used as a switch, a threshold level of current reached on the base pin makes the other two pins conduct (switch on), and a current below the threshold level makes the other two pins stop conducting (switch off). Often regular transistors are assembled back to back in something called a Darlington Array Pair; this makes transistors more like digital switches and less like gradual amplifiers.

Various types of transistors are used with robots. Basically, you have to look up the spec sheet to find the part numbers and what each part can do. Spec sheets can be found on the manufacturer's Web site or in a data book. A transistor known as a MOSFET transistor, for example, controls motor speed and direction. Smaller transistors such as a 2N2222 are used for switching LEDs on or off.

Integrated circuits

If you're the type who likes his or her electronics prepackaged, you'll love ICs. *Integrated circuits* (ICs) consist of several transistors, resistors, and sometimes other components, all integrated into a tiny circuit and usually packaged onto a plastic or ceramic box with pins called a chip. ICs (see Figure 5-10) come in different varieties, including CPUs, logic chips, amplifier chips, gate array logic chips, and memory chips.

Although building a mammoth circuit from scratch gives you a better understanding of the circuit and electronic principles, it's always a better idea to use ICs to simplify your life and reduce your robot's weight.

A common integrated circuit is the 7400 series of chip. Some 7400 series chips used for robotics are the 7404 and the 7414 inverter/buffer chip, both used for drawing enough amperage to a chip to power LEDs. Another common 7400 series chip you can use is the 16-pin 74138 chip to select one of eight devices.

To find other 7400 series chips useful for robotics, consult a 7400 specification manual, which you can find by visiting electronics manufacturers' Web sites, such as www.jameco.com. The *low-power shottkey* (LS) or *advanced lower-power shottkey* (ALS) versions are the most common and efficient versions. You can spot these by looking for an LS or ALS embedded in the number, such as 74LS14.

Figure 5-10:
Integrated
circuit
chips.

MCUs

Microcontroller units (MCUs) are very large ICs fashioned into tiny computers
with program memory, data memory, and input and output (I/O) lines integrated into one neat package. Unlike MCUs, *central processing units* (CPUs)
don't have I/O lines integrated on the chip, so they can't control things
directly. To get a CPU to turn an LED on or off, you would have to tap into the
memory and address lines and create an I/O memory map. Trust me, this is a
major pain. That's why MCUs are great for robotics and why CPUs generally
aren't. MCUs are useful for robotics also because they fit on a tiny computer
board, use very little power, and can control devices such as LEDs and
motors directly through their I/O lines.

Some common MCUs used in robotics are the 8051, HC11, and PICs (programmable ICs). These days, robot builders even use computer chips powerful
enough to be used in a personal computer.

Creating circuit boards

A *circuit board* is a board with an electronic circuit printed on it in copper.
The board has holes that you use to attach electronic components and chips
to run your robot. Your robot's brain is usually made of a circuit board with a
computer chip and other components, which you attach to your robot chassis.

Robot boards such as the Basic Stamp 2, the Handy Board, and the PC104 architecture are rapidly becoming the standard for robotic brains. However, these boards don't always have every feature (for example, sonar ranging) that you want. In some cases, you will need to add functionality to your robot by creating your own printed circuit boards. Following is information on a few ways to create circuit boards.

Breadboards

The easiest way to put together an electronic circuit board is to prototype one first using a breadboard like the one shown in Figure 5-11. A *prototype* is just a fancy name for an experimental circuit. A *breadboard* is a plastic sheet with holes connected by wire strips. You plug in a chip, plug wires into vertically adjacent holes, and then route the wires to another part of the board, thereby creating a circuit. You can buy breadboards at your local Radio Shack or through an electronics dealer such as Jameco (www. jameco.com).

Breadboards are great for prototyping and testing, they're not permanent because the wires can come loose and fall out. If you want a stronger, permanent solution, use a perf board, described next.

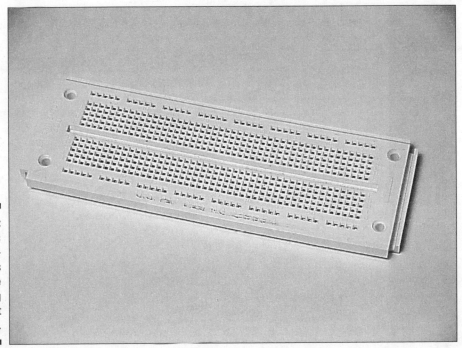

Figure 5-11:
Plastic bread-boards make prototyping a circuit easy.

Perf boards

If you've already prototyped your circuit and are sure it works, you may want to solder a perf board (also called a prototype board or a proto board). A *perf board* is a circuit board predrilled for electronic components and chips. A regular gridwork of traces are provided to allow you to wire your circuit. With perf boards, you have to solder parts as well as jumper wires to make your circuits. Figure 5-12 shows the top side and the bottom of a typical perf board.

Don't be confused with the term *perf board*. A perf board is a printed circuit board. But instead of having traces already laid out, a perf board has a regular grid of traces that you can use to build many different circuits.

You can purchase preprinted and drilled perf boards from Radio Shack or just about any another electronics parts source.

Printed circuit boards

Most *printed circuit boards*, or PCBs, are made for a specific purpose. That means they have the specific holes and traces needed to create a particular circuit. Perf boards are kind of like PCBs, except they have just a grid of holes so that you can plug in a chip or component wherever you want; you have to make your own circuit by soldering wires. (That is, there's usually no traces to speak of, just a grid of holes.)

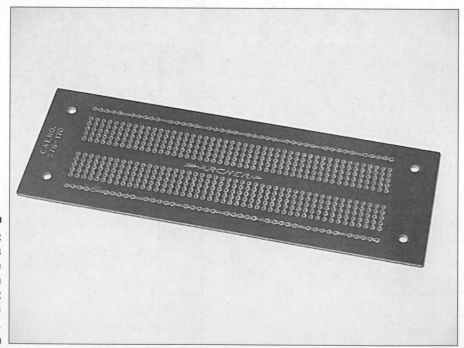

Figure 5-12:
Perf boards let you wire a custom circuit that will be permanent.

Custom-printed circuit boards

If you've made a useful PCB that you think others may want to have, you might want to either hand them out to friends or sell them. In that case, you could do a production run of your board, having your circuit boards professionally made by a PCB manufacturer.

PCB manufacturers such as Alberta Printed Circuits (www.apcircuits.com) or Proto Express (www.protoexpress.com) often allow you to order just a few PCBs inexpensively. (Other larger PCB manufacturers have minimum orders of 1000 boards.) Although professionally made PCBs are costly for small runs, they ensure a good circuit board.

Layout software allows you to create a computer-aided draft of your electronic circuit. With the software, you can lay chips down, connect pins with virtual traces, and place pads and text. After you finish your design with the layout software, you output the design to a file suitable to send to a PCB manufacturer who then builds the circuit. Using layout software can help you make your design more efficient and compressed to reduce cost.

Another way to prototype boards is to etch the boards yourself, like the one shown in Figure 5-13. The most common way to make your own boards is using the photo-etch process. You can get a kit to make homemade boards from Radio Shack or MG Chemicals (www.mgchemicals.com). Etching your own boards can give you professional results and can be cheaper than buying from a prototype company. Etching also involves a lot of work involved and a big startup fee, but once you get going, you can turn around a board in minutes.

Figure 5-13: A proto-typed printed circuit board.

Off-the-shelf boards

You can find circuit boards to handle all sorts of functions, such as motor control, serial servo control, sonar control, video camera control, speech synthesis, and digital compass measurements. Before you attempt to design, say, a motor controller from scratch, you may want to see whether you can purchase it off the shelf and incorporate it into your robot design. You can find off-the-shelf boards at Jameco (www.jameco.com), Parallax (www.parallaxinc.com), and the Robot Store (www.robotstore.com).

Power to the Robots

Batteries and other power sources for a robot are like the engine in your car. You and your robot will not get far without them. In addition to affecting your robot's performance, batteries and other power sources affect your bank account.

ARobot, which you build in Part III, comes with a plastic holder that carries eight AA-type batteries. This produces approximately 12 volts when using alkaline batteries and less when using rechargeable batteries. For an important discussion of how to save money when you buy batteries, see the "Juicing it up with batteries" section in Chapter 3.

Everyone knows what a battery is but you may not know everything that happens inside that little casing to produce energy. Because batteries play a big part in the world of robots, I'd like to mention a few things you may or may not know about them:

✔ All batteries have a voltage and current rating. Voltage and current are like air in a pipe. The voltage is the pressure of the air, and the current is the volume of air moving through. If the pipe is capped off, there is no airflow. In the same way, a battery that is not powering a device has no current flowing.

✔ Most household batteries have a voltage rating of about 1.2 to 1.5 volts per cell and can supply several hundred milliamps of current for an hour or more. The less current that's drawn out of the battery, the longer it will last.

✔ Because most products need more than 1.5 volts to operate, they use several batteries in series. For example, if you have four 1.5-volt batteries in series, you'll get about 6 volts. If you need more current but not more voltage, you can attach the batteries in parallel, which causes the current to increase while keeping the voltage the same as that of each individual cell. Batteries in series sum the voltage. Batteries in parallel sum the amperage.

Series and parallel demystified

Two terms that you'll run across often when dealing with electronic components — especially batteries, resistors, and capacitors — are series and parallel.

A *series* connection is made by connecting components end to end. For example, to connect three batteries in series, you connect the positive terminal of one battery to the negative of another. When you connect batteries this way, you add the voltage. Most appliances and toys have battery compartments made to connect them in series. (You can check for yourself.)

To connect batteries in *parallel,* you connect all the positive terminals of the group together, and then connect all the negative terminals together. When connected in parallel, the battery voltage does not change, but the current capacity is increased.

Bench DC power supply

Much of your work as a robot guru will involve sitting at your computer pounding those keys writing cool code to get your robot to do stuff. In the process, you'll be downloading programs to the robot and checking to see whether everything is working. This repetitive process can take its toll on batteries.

Enter the bench DC power supply, shown in Figure 5-14. It won't write code for you, but it will help save your batteries by providing an alternate power supply. Here's how this works. ARobot can operate on 10 to 16 volts. Lucky for you, this just happens to be the voltage needed to run car stereos, CB radios, and other 12-volt DC appliances. Because of this, 12-volt DC power supplies are readily available from electronics distributors such as Radio Shack.

To use a bench DC power supply instead of batteries, you have to make a cable with a connector that's just like the battery cable connector. Use awg (American wire gauge) 10 wire or larger and make the cable as long as needed — 25 feet would be fine. A long cable will allow you to test the robot (for example, letting it move across the room and back) while using the power supply.

Figure 5-14:
Using a
power
supply
to save
battery life.

You can get molex connectors that you can put on wires from your battery, robot, and power supply. Then you can unplug your battery from your robot and plug in your power supply. You could even use a DC adapter to plug into a DC plug on your robot. (Choose a plug that automatically disconnects your batteries from the circuit.)

When ARobot's ready to slip its bench power leash and go roaming, just unplug the power supply and plug in the battery pack.

Because ARobot draws less than 1 amp of current, a very small power supply will do. A 12-volt, 3-amp power supply should cost you less than $30.

You may have noticed a trend in power supply polarity reflected in the use of battery wire colors: Red is positive and black is negative (no voltage or ground). These colors are almost universally recognized — you'll even see them used on automotive jumper cables and on your voltmeter. The DC power supply mentioned in this section also has a color-coded binding post to help prevent unpleasant polarity mix-ups.

Tethered power

Some robots, such as walking robots, can't afford to haul around heavy batteries. If a robot is designed to stay in one room and a power cord can reach it at all times, it may be better to use a tethered power cord. Tethers limit a robot's range and can sometimes get in the way, but they also give your robot unlimited power. Sometimes both power and data are tethered through a cable back to a main computer.

Fuel cells on the horizon

Fuel cells are a set of plates that convert a fuel such as hydrogen into electricity. They promise longer and stronger power output than batteries do. Fuel cells come in various shapes and sizes, from small ones that fit in cell phones to big ones used in automobiles. However, fuel cells aren't yet widely available on the market for practical robotic use.

Chapter 6

A Programming Primer

I promised you at the beginning of this book that you didn't have to be a programmer or an engineer to get into robotics. However, for the remainder of this book, I *do* include code snippets all over the place. These snippets are designed to work with ARobot, which you build in Part III, but the general ideas behind them can be ported to other robots. Feel free to use this code to make your robot do all sorts of cool tricks and behaviors.

I'm not trying to thoroughly educate you on how to program. I give you just enough information to deal with the simple programming required in this book and to get the gist of what the code listings in this book do.

Programming isn't tremendously difficult. If you want to discover more about it, I suggest that you refer to books written on various programming languages (of course, I recommend those in the *For Dummies* series from Wiley Publishing, Inc.) for a more thorough treatment.

Software and Computers

Robots and computers go hand in hand (or hand in mouse, as the case may be). And where there's a computer, there's programming.

Kits such as the ones you'll work with in this book keep programming on your part to a minimum. If you want to add functionality to your robot, however, you'll have to know something about programming. The more you know, the more you can make your robot do.

If you can't imagine yourself doing even a bit of programming, use a prepro-grammed robot that requires no programming. And certain robotics kits, such as Lego MindStorms, have their own easy-to-learn software languages.

I give you the programming details you need to get the code (which I provide for you) into your robot and to understand the gist of what that code will make your robot do.

What follows are a few concepts that you should know about how I approach programming a robot.

Keep it modular

Programs are made up of code — lines of text that tell your robot what to do. One good programming discipline is to make your code modular. *Modular programming* means creating reusable and easily modifiable blocks of code that may be added or removed. Tasks in modular programming are broken up into these logical function blocks of code.

The opposite of modular code is spaghetti code. Spaghetti code is so inte-grated and hard to follow that it can't be modified or removed easily. (It's called *spaghetti code* because if you were to draw lines to follow the logic of the code, you'd end up with a page that looked like a plate of spaghetti.)

Modular code usually has a method for sending and receiving data, but the details of its internal operation are hidden from view. If you write modular code, you save yourself a lot of time both when writing and *debugging* (locate and correct problems with) your robot's program.

I generally divide programs into two categories: high level and low level. The low-level software module handles and regulates hardware access, such as turning on motors or reading sensors. The high-level software controls the task at hand, such as navigating down a hallway.

Programming: The low road

Low-level programming involves developing the software that controls robot hardware directly through what is called a *device driver*. You've probably encountered device drivers before because they also control hardware peripherals such as printers and modems on your desktop computer.

In addition to controlling hardware, low-level software might also be written to provide operating-system-like functions, such as grouping and controlling device drivers, or software application functions such as error checking.

Low-level programming usually involves simple commands such as turning a motor on or off. Low-level programs access the hardware directly but often hide the details of the hardware from higher-level programs. This modularity makes life simpler if you have to hunt down a problem with your software.

Rarely do low-level programs access high-level functions, except to interrupt the high-level software in case of an emergency that might require the robot to shut down.

Programming: The high road

High-level programming usually involves developing the software application that will ultimately control your robot. High-level programs usually perform the decision-making processes of the robot.

High-level software should never access hardware directly, though it may direct low-level software to do things. For example, if you're making a fire-fighting robot, your high-level program will search for fire and snuff it out, utilizing the device drivers to control or take readings from heat sensors.

In this book, you'll encounter a few simple algorithms that high-level software might typically run. You'll also explore a few techniques used by high-level programs to decide what to do and to issue commands to the operating system and device driver software.

Programming quick-start

Many types of programming languages are used for computers. Some of the most popular are C, Basic, Cobol, and assembly language. Of the many variations of these languages floating around, each has its own strengths and weaknesses. The language used in the BS2 is PBasic, which is based on the popular Basic language that many of the first personal computers used. PBasic instructions are easy to understand and provide all the power needed for a small processor, like the BS2 (Basic Stamp 2).

In a nutshell, a *program* is just a series of instructions that tells the computer — in this case, the BS2 controller — what to do. The program resides in non-volatile memory in the BS2. The BS2 executes the commands in the program one at a time, in sequence. For the most part, you can simply take chunks of code in this book and enter them into the Basic Stamp editor.

When you download a program from your computer to the BS2 controller, the Basic Stamp editor converts the program text into small instruction packets, called *tokens*. The BS2 stores these tokens and interprets them while the program is running. A single token might result in many smaller instructions being executed.

Programming resources

Complete information about programming in PBasic is available from the Parallax Web site at www.parallax.com. Check out the multimedia tutorial, the Basic Stamp Editor User's Guide with programming references, and several application notes. These are all free for the taking. Note, however, that the programming documentation is long (several hundred pages), so you're better off buying a book if you're serious about programming.

Any program that you download to the BS2 controller gives you complete control over all of your robot's features. For example, you can read the status of the whisker sensors, turn the steering left or right, turn LEDs on or off, make sounds with the speaker, read the status of the switches and jumpers, and control devices connected through the expansion port. You can do almost anything that your little heart desires with your robot, providing that enough memory exists to hold the program.

Anatomy of a Program

You won't become fluent in programming during the course of building your robot, any more than spending a weekend in France leaves you speaking like a Parisian. But you'll feel a lot smarter if you can figure out the basics of what the programs in the book do before you download them into your robot. That's what this section is all about.

Programs consist of things such as comments, program code, and variable declarations. The arrangement of these pieces and the format that you use can make life a lot easier when you need to debug the program or when others have to work with your program.

The most important thing is that your program works. How your program looks, however, tells folks a lot about your ability, logic, and organization skills.

Comments

Every program should have comments describing what the program is doing as well as why and how. Without comments, a complex program can be impossible to debug or to edit by anyone but the author. In fact, after a few months, you might not even be able to make sense of your own code. Adding a healthy dose of comments is a good habit.

The number of comments can often greatly outweigh the amount of actual program code!

The first few lines of a program should contain information about the programmer and the program. Make sure that each program you write begins with this information:

✔ Program name

✔ Program function

✔ Programmer's name

✔ Date

✔ Information on how to use the program

Here's an example:

```
'test1s.bs2 - Test LEDs and speaker - Roger Arrick 2/22/03
'This program flashes both LEDs and beeps the speaker.
```

Declarations

The next section of your program *declares* variables, constants, and processor resources. This declaration is required so that the computer will know how to work with these items. What does this list of programming jargon mean in English?

✔ **Variables:** Named locations in memory that are used to store values while the program runs. Variables can be referred to and used by many instructions (such as read or write).

✔ **Constants:** Fixed numbers that you name in the code. This can make it easier to understand the program logic and to make changes because the number is defined in one location and labeled.

✔ **Processor resources:** Port pin numbers and other on-chip hardware devices. These ports and devices are usually connected to external hardware, and you should identify them for readability.

Here's the declaration section of the sample ARobot program:

```
'sample.bs2 - Sample ARobot program - Roger Arrick 2/22/03
'
net      con 8      'coprocessor network pin
Baud     con 396    'coprocessor network baud rate
```

This part of a program contains only constants, which are used to set parameters needed to communicate with ARobot's coprocessor.

Here's the declaration from another program:

```
'Test22.bs2 - Sample ARobot program - Roger Arrick 2/22/03
'
RedLED          con 10       'Red LED port pin #
Distance        var byte     'Variable for distance to travel
```

In this code, RedLED is a constant and Distance is a variable.

Instructions

The rest of a program contains the instructions that make things happen. The types of instruction can be broken down into decision-making statements (if-then); input/output commands (high, low, serin); memory access (data, read, write); math operations (+, –, *); looping (for, next); and so on.

You don't have to commit all this to memory. Instead, use this book as a reference.

Sample program

Listing 6-1 is a simple program to show you how all the pieces I've described come together to form a complete application. This simple program flashes the LEDs and moves the steering motor in response to the whiskers. The embedded comments describe each line. *Note:* PBasic is case-insensitive, so you can use uppercase and lowercase letters interchangeably.

See whether you can follow the flow and predict whether this program will operate.

Listing 6-1:   A Sample Program

```
'sample.bs2 - Sample ARobot program - Roger Arrick 2/22/03
'The LEDs flash and the steering motor moves according
'to the whiskers

net        con 8        'coprocessor network pin
Baud       con 396      'coprocessor network baud rate

Start:     freqout 9,200,1500        'Beep
           low 9

s1:        if in0=0 then wh1        'Left whisker?
           if in1=0 then wh2        'Right whisker?
```

```
                serout net,baud,["!1R180"]  'Steer center
                pause 500                   'Pause 1/2 second

                low 10                      'Red LED on
                high 11                     'Green LED off

                pause 500                   'Pause 1/2 second

                high 10                     'Red LED off
                low 11                      'Green LED on

                goto s1                     'Loop

    wh1:        serout net,baud,["!1R1FF"]  'Steer left
                pause 500                   'Pause 1/2 second
                goto s1

    wh2:        serout net,baud,["!1R101"]  'Steer right
                pause 500                   'Pause 1/2 second
                goto s1

    ' End of program.
```

 Notice in the program in Listing 6-1 that the steering is performed by a `serout` command. This command simply sends data from the BS2 to the coprocessor that handles the steering motor and the drive motor. The coprocessor relieves the BS2 from having to handle these functions itself.

Useful Code Snippets

If you read the newspaper on a regular basis, you recognize that the media uses certain phrases all the time: *Sources close to the President report* or *The identity of the victim is being withheld,* for example. Writing code is kind of like writing a news story: Certain snippets show up again and again, and having them in your programming vocabulary makes each story — I mean, program — easier to write.

The fact is that when you're writing a program for your robot, you'll find that many actions are performed over and over, and they pop up in almost every program. These are often simple actions such as turning the steering mechanism, moving forward, turning on an LED, or making a beep sound through the speaker.

Here are all the common things that appear repeatedly in the code samples in this book. Notice the useful comments at the end of each line of code that help you identify what the heck is going on.

Controlling LEDs:

```
low 10                    'Red LED on
high 10                   'Red LED off

low 11                    'Green LED on
high 11                   'Green LED off
```

Controlling the speaker:

```
low 9                     'Speaker off

freqout 9,200,1500        'Beep (high pitch)
low 9                     'Speaker off

freqout 9,200,400         'Beep (low pitch)
low 9                     'Speaker off
```

Reading the SW1 and SW2 pushbuttons:

```
if in14=0 then wh1        'If SW1 on then goto wh1
if in14=1 then wh1        'If SW1 off then goto wh1

if in15=0 then wh1        'If SW2 on then goto wh1
if in15=1 then wh1        'If SW2 off then goto wh1
```

Reading whiskers:

```
if in0=0 then wh1         'If left on then goto wh1
if in0=1 then wh1         'If left off then goto wh1

if in1=0 then wh1         'If right on then goto wh1
if in1=1 then wh1         'If right off then goto wh1
```

Reading the J6 and J7 jumpers:

```
if in12=0 then wh1        'If J7 on then goto wh1
if in12=1 then wh1        'If J7 off then goto wh1

if in13=0 then wh1        'If J6 on then goto wh1
if in13=1 then wh1        'If J6 off then goto wh1
```

Controlling the steering:

```
'These require a variable named charn

serout 8,396,["!1R180"]   'Steer center
serin 8,396,[charn]       'Get response

serout 8,396,["!1R1FF"]   'Steer left
serin 8,396,[charn]       'Get response
```

```
serout 8,396,["!1R101"] 'Steer right
serin 8,396,[charn]      'Get response

serout 8,396,["!1R100"] 'Relax steering
serin 8,396,[charn]      'Get response
```

Controlling the drive motor:

```
'These require a variable named charn

serout 8,396,["!1M1160028"] 'Forward 10"
serin 8,396,[charn]            'Get response

serout 8,396,["!1M1060028"] 'Reverse 10"
serin 8,396,[charn]            'Get response

serout 8,396,["!1M116FFFF"] 'Forward forever
serin 8,396,[charn]            'Get response

serout 8,396,["!1M1100001"] 'Drive motor off
serin 8,396,[charn]            'Get response
```

Useful Subroutines

Subroutines are (usually) short chunks of code that collectively serve a specific purpose and are used repeatedly in a program. Putting code in a subroutine eliminates the need for multiple copies of the same item in your code, saves memory, and makes changes to the code easier.

The following subroutine flashes the red LED three times. To call the subroutine, you use the gosub FlashRL command.

```
FlashRL:
For I = 1 to 3
  low 10                    'Red LED on
  pause 300
  high 10                   'Red LED off
  pause 300
next
return
```

Creating a program by starting with existing, tested subroutines is much easier than creating them from scratch.

You use the gosub command in your code to begin a subroutine (among computer geeks, this is known as *invoking* the command), followed by the name of the subroutine. When the program runs, that command will run, or *execute,* the subroutine.

You can use the upcoming subroutines to begin building your programming inventory.

Starting a program

The subroutine in Listing 6-2 starts a typical program. It begins by centering the steering, then it causes the robot to beep twice, and finally it flashes the red LED. The subroutine finishes only when you press and release SW1. (If you don't know where SW1 is located, check out the first figure in Chapter 9, which labels all the elements on the controller board.) Use this subroutine at the beginning of any program to get things cleaned up (such as straightening the steering and turning the LEDs off) and to alert your robot to await impending user input.

Listing 6-2:   Starting a Program

```
' The start subroutine
'Start a program with this routine
'Centers steering, beeps twice, waits for SW1 pressed
'
Start:      high 11                    'Green LED off
            serout 8,396,["!1R180"]    'Steer center
            freqout 9,150,2000         'Beep twice
            pause 20
            freqout 9,150,2000         'Beep twice
            low 9                      'Speaker off
            serout 8,396,["!1R100"]    'Relax steering
start1:     high 10                    'Red LED off
            pause 100
            if in14=0 then startd      'If SW1 then startd
            low 10                     'Red LED on
            pause 100
            goto start1                'Loop
startd:     if in14=0 then startd      'Wait till SW1 off
            return                     'Finished
```

Backing away to the left

Cats use whiskers to sense their environment, and so does your robot. Whiskers are commonly used in a program to determine whether the robot has run into something. If it has, you'll want your mechanical friend to back away from the object in the direction determined by the whisker (left or right). If the robot's right whisker detects a collision, for example, the robot backs away to the left (in the opposite direction).

Listing 6-3 is a short subroutine that responds to a right whisker. It stops the drive motor and then moves the robot in reverse, away from the object.

When the subroutine returns, your main program can tell your robot to resume forward motion. By changing a few lines of code, you can make a subroutine that tells the robot to move away to the right.

Listing 6-3:   Backing Away to the Left

```
' The back1 subroutine
'Move the robot back and to the left to avoid an object.
'
back1:      serout 8,396,["!1M1100001"]   'Drive motor off
            pause 300
            serout 8,396,["!1R101"]       'Steer right
            pause 500
            serout 8,396,["!1M1060028"]   'Reverse 10"
            pause 1500
            serout 8,396,["!1M1100001"]    'Drive motor off
            pause 300
            serout 8,396,["!1R180"]        'Steer center
            pause 300
            return
```

Sounding an alarm

It's a dark and stormy night. Suddenly an intruder breaks in and steps on your robot. This is serious — somebody (or something) should raise an alarm. Your robot can handle this one for you.

The subroutine shown in Listing 6-4 can be used to inform you of a panic situation. The robot's LEDs flash, and then the speaker sounds an alarm (the equivalent of the robot shouting, "Hey, somebody just stepped on me!"). The alarm lasts about four seconds and can be called multiple times if you want it to go on and on.

Listing 6-4:   Turning on the Alarm and Flashing Lights

```
' The panic subroutine
panicv      var  byte                    'Variable for looping

'Panic - flash the LEDs and sound the alarm
'
panic:      for panicv = 1 to 10         'Loop 10 times
            low 10                       'Red LED on
            high 11                      'Green LED off

            freqout 9,100,1500           'High pitch sound
            pause 100
```

(continued)

Listing 6-4 *(continued)*

```
high 10              'Red LED off
low 11               'Green LED on

freqout 9,100,500    'Low pitch sound
pause 100

next                 'Loop until done
return               'Done
```

By using this and other subroutines, you make your robot programming a much easier task and build a neat library of useful code you can reuse again and again.

Okay, I Programmed . . . Now What?

Many advanced robots that you can build require that you develop the robot software on a PC. In that case, you need a way to get that software from your PC into your robot. That's where downloading comes in.

Each robot may connect to your PC in a different way, but generally software is downloaded through some sort of RS232 or USB type cable that you connect to both your PC and to your robot. When you purchase a robot kit that uses such a cable, you need to consult the user manual on how to download software for your particular robot.

To download software to ARobot — the robot kit you work with in this book — you use a common computer controller called the Basic Stamp 2 and a cable that connects a PC to the robot. Chapter 10 gives you a good overview of the downloading process.

Part III
Building a Programmable Robot

The 5th Wave   By Rich Tennant

Ever since I installed the "Dork Sensor," it won't leave my sister and her boyfriend alone.

In this part . . .

This part is like Christmas morning, because this is where you finally get to unwrap all the robot para-phernalia and start assembling things. (And just like that electronic gift you get every year from Aunt Tillie, you'll even need batteries.)

By working through the four chapters in this part, you'll move from preparing the parts of a robot kit, to assembling the chassis, wheels, and sensor whiskers. You find out how to use the editor software that enables you to program your robot and get your computer and robot connected.

The only thing missing is the tree, the Christmas carols, the snow . . . well, just about everything. But that Christmas-morning excitement as you see your robot take shape will be abundant.

Chapter 7

Preparing to Build a Programmable Robot

. . . .

. . . .

*T*he world of robotics holds some murky gray areas, including much debate about what type of machine can even be called a robot. In my mind, to be a true robot, a machine must perform on its own certain semi-intelligent functions: sense its environment, make decisions, and perform some actions.

You might have noticed the availability of robots in catalogs and in toy stores, but you might not realize that most of them are simply remotely-controlled vehicles. They can't make decisions and perform intelligent actions — the very capabilities that I maintain define a robot. Instead, the operator of these machines initiates their actions with buttons and switches.

In my definition of robot, simple remotely-controlled cars wouldn't make the cut, but a remotely-controlled planetary explorer that can make some decisions on its own would. (But remember, even the most intelligent robots get their ultimate marching orders from you, their creator.) One key attribute that makes for a more intelligent robot is that it's programmable — that is, you can download programs to the robot that allow it to carry out processes and make decisions.

They say you should walk before you run, and walking the walk with a robot kit is a good way to ease into robotics. Starting with a robot kit such as ARobot or Soccer Jr. is a great way to get to know the various pieces of hardware and procedures used to build almost any robot. By investing only a few hours of your time, you end up with your first robot.

This is the first chapter in a set of four in which you build a powerful, programmable robot called ARobot (pronounced "a robot"). You can see exactly what you'll end up with in Figure 7-1.

Figure 7-1:
ARobot:
A real
program-
mable robot.

Building ARobot requires a medium skill level, which means you should have no trouble putting it together if you can build a model car or airplane. The kit includes all the hardware necessary to build the robot, and it usually takes about two hours to complete.

Robot Components Overview

Before you grab a screwdriver and begin building, however, it helps to have a general understanding of the components that make up the body of your robot as well as the more specialized parts that control its actions.

Looking at basic robot parts

Robots consist of a few basic components:

- A **base,** which is like the robot's body, to which you attach its various parts
- A set of **wheels** that your robot uses to roll around the room
- A **steering mechanism** that helps the robot direct those wheels

✔ A **drive motor** that makes the robot move

✔ A **battery pack** that gives juice to the robot so it can get up and go

✔ **Cables** to connect various components together

✔ A **controller** circuit board that acts as the robot's brain

Building the basic robot involves attaching the wheels, drive motor, and the steering mechanism to the body, adding the controller and a battery pack, and connecting some cables.

You've probably already seen parts such as wheels and a battery pack. An unfamiliar robot feature, however, might be the controller circuit board (see Figure 7-2). For a robot to perform sensing and decision-making functions, a computer must be involved. In robot lingo, that computer is a *controller*.

Controllers, which are simply computers that have specialized features for robots, usually have more input/output (I/O) ports than the personal computer sitting on a desk. Controllers need these I/O ports to read sensors (such as a sensor that picks up data about temperature or light) and control actuators such as motors and speakers. When you've assembled all the parts of your robot, your last steps will be to install the controller board and controller chip and connect the robot to your computer. Then you're ready to download programs that tell the robot what to do.

Figure 7-2:
A typical controller board and chip for robot building.

ARobot is large enough and powerful enough to accommodate many accessories. You explore these in Part IV.

Looking at ARobot's components

First things first. Open the kit box and see what bits and pieces it contains. Lay everything out around you and open the assembly book — but don't remove the parts from their bags yet. Figure 7-3 shows you what the ARobot kit stuff looks like.

You'll find the following items:

- Dual, front whisker sensors that are used to detect objects that the robot encounters
- Rear-wheel steering mechanism with an RC (remote-control) servo motor
- Bidirectional, front-drive gear motor (also called a drive motor) that moves the robot
- Optical encoder that your robot uses to measure the distance traveled when it goes for a stroll
- Metal body components and brackets
- Wheels
- Drive and steering motors
- Battery pack and cable
- Cables, which carry signals between the controller and components
- Controller board used to hold the Basic Stamp 2 controller chip, which provides the brain for the robot

Note: The Basic Stamp 2 controller chip is required for operation and is sold separately.

Beware of small parts

ARobot doesn't contain any dangerous voltages, radioactive Kryptonite, or blinding lasers, so you don't have to be concerned with a whole slew of safety issues. Before you get started, however, remember that small children and small parts don't mix.

Also consider putting all your small parts in separate paper cups, plastic bags, or some other containers so that the parts don't roll off the table, into the waiting mouth of the vacuum cleaner or those little ones.

Figure 7-3:
The parts of
your kit and
the tools
that you
need for
assembly.

In addition to these major items mentioned, the kit also contains a bunch of screws, spacers, washers, brackets, and other hardware for putting ARobot together. (If the King's men had had all this stuff, they could have put Humpty Dumpty together again in a heartbeat.)

You might notice some extra parts such as screws and nuts. Don't panic. The manufacturer puts extra parts in there just in case you lose one or they make a mistake. I suggest you put all the extra parts in a bag and keep them for a while just in case. You could also make any extra hardware part of your robot parts inventory.

Assembly Process Overview

Before you jump right into assembling all this stuff, I thought I'd give you an overview of the steps of the building process. Building ARobot involves performing the following steps:

1. Organize and prepare ARobot's parts.

2. Sand, paint, and decorate ARobot (optional).

3. Connect the whiskers.

4. Install the front wheel assembly with axle and drive motor.

5. Mount the encoder sensor.

6. Assemble the steering motor and rear wheels.

7. Mount the controller board.

8. Install the Basic Stamp 2 controller (the computer that controls ARobot) on the controller board.

9. Mount the battery holder.

10. Use the body cable and other cables to connect various components.

11. Route cables to tidy things up.

12. Add the Basic Stamp 2 controller chip to the controller board.

Robot building isn't difficult, but it does involve several steps. To break it all up into manageable chunks, you'll complete Steps 1 and 2 in this chapter. The rest of the building process is covered in Chapters 8 through 10.

How long it takes you to complete a project depends on your speed, the number of interruptions, and any trouble you may encounter. I suggest beginning only when you can commit at least one hour of time.

During this process, you'll be utilizing basic assembly and wiring skills. You'll also need a basic knowledge of electronics terminology. You can brush up on these skills and concepts if you're feeling rusty by reviewing Chapters 4 and 5.

Organizing and Prepping the Parts

Okay, it's time for all you robot builder wannabes to put your money where your mouth is and build your first robot. When you do, you'll be permanently sucked into the world of robot lovers and never be able to find your way back to the simple life you used to lead. But trust me — the joy of building robots is worth it all!

Taking it one step at a time

The first things a successful robot builder needs are patience and a willingness to learn as you go. Master basic skills before tackling a complex robot. Discovering simple mistakes now can help you avoid costly mistakes later. Pay attention to the small stuff, and the small stuff will take care of you and your robot.

Now's the time to decide whether you really want to build this thing because you can't return a half-baked robot.

Gathering your tools

If you have nightmares of having to assemble an arsenal of high-tech, sci-fi-type tools to work with robots, let me reassure you that the only tools you need are common hand tools found in most kitchen drawers (far in the back, behind the out-of-date coupons for yogurt):

- Phillips screwdriver
- Needle-nosed pliers
- Hobby knife

In addition to these basic tools, you also need the following:

- **The Basic Stamp 2 (BS2) controller:** The BS2 makes ARobot think like a real robot.
- **Sandpaper, paint, a brush, a sloppy shirt, and some extra time:** These items are required only if you plan to paint your robot.
- **A fresh supply of eight AA size batteries:** You're usually better off buying the best batteries because they last so much longer. See the section titled "Juicing it up with batteries," in Chapter 3, for more battery advice.

The faint of heart will be happy to hear that soldering isn't required to complete the ARobot kit because the controller board is presoldered and tested, as are the body cable, motors, and other cables.

Preparing the parts

One of the first jobs you have to tackle is cleaning up the robot's body, or *chassis*. ARobot's body parts (a base, two motor brackets for the front wheel, and an encoder wheel) are made of aluminum and have been precisely cut with an expensive laser. However, other than bending and some mild sanding, these metal pieces have been given no other preparation.

Some metal edges have *burrs* and *slag* (metal jagged thingies). To take care of these imperfections, use a metal file or some 220-grit sandpaper on those rough edges until you have a nice smooth finish (see Figure 7-4). If necessary, use a hobby knife to trim excess material — always cutting away from your fingers and other body parts, of course.

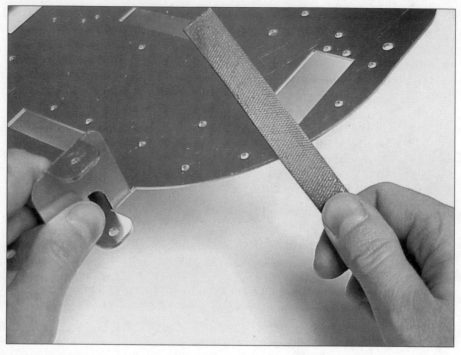

Figure 7-4:
Filing
produces
smooth
edges.

To Paint or Not to Paint?

This is a good time to decide whether you want to apply paint to ARobot, because it's the easiest time to get at all the parts before they are assembled and harder to reach.

The rugged look

Many people choose to leave the metal pieces unfinished because that approach is simpler, provides a hip, industrial look to the robot, and allows you to get on with construction without waiting for paint to dry.

If you're going with this minimalist approach, at least use sandpaper on the metal surface to give it a brushed look. You can use fine-grit sandpaper, buffing pads, or even steel wool. Make sure that you clean the surface before construction, though, because small metal particles can ruin a circuit board's entire day.

Playing dress up

Why shouldn't your robot have a wardrobe? A quick visit to the local hobby store will give you some ideas about dressing up your robot. For example, decals made for model airplanes and cars will work perfectly when stuck on his little chassis. Most of these decals are easy to apply because they use an adhesive backing that sticks to most surfaces.

Or use other paint colors and paint a mural on your robot — the planet from *Lost in Space* comes to mind. Your only limits are your imagination, time, and budget.

The snazzy look

Paint can definitely give your robot a colorful personality; you can even match the décor of your house. Choose a paint that's made for metal surfaces. At home improvement, hardware, or craft stores, you can find many interesting metal paint colors — everything from flat black to hot pink — as well as specialty paints that produce stone-like textures or crackled surfaces. Figure 7-5 shows a freshly painted chassis, ready to be assembled.

Figure 7-5:
Paint is your robot's skin.

Lightly sand the flat areas to give paint a good surface to adhere to. Be sure to clean off any metal dust or debris before applying the paint.

Although spray paint should stick to the metal surface of ARobot's body parts, I recommend applying a primer first. A *primer* is simply a special paint (usually gray) that's designed to stick to the metal and give the paint a good surface to adhere to. Spray primer is very inexpensive; find it along with the other spray paints at your local hardware store.

Here are a few helpful hints to remember when using spray paint:

- ✔ Follow the instructions on the product.
- ✔ Paint in a spot that has good ventilation, preferably outside.
- ✔ Avoid painting near sources of heat (fireplace, heaters, and so on).
- ✔ Paint far away from cars, concrete, and anything else you don't want to turn bright blue (or pink or green).
- ✔ Use several light coats instead of a single thick coat for a smoother finish.
- ✔ Allow plenty of time for the paint to dry — preferably overnight.

Now that you've prepped and primped your robot (or even if you decided not to gaudy it up with paint), the next step is to start assembling the thing. Chapter 8 is your next stop on the road to actually having a robot to call your own.

Free Rear Whisker Kit

ARobot's rear whisker kit includes all the hardware and documentation needed to add 2 more whiskers. Usually folks place them on the rear of the robot, but you can also place them on the sides or add them to the front to indicate the position of obstacles more accurately.

This offer is only valid for customers in the USA.

To Receive your free ARobot rear whisker kit:

■ Supply us with 1 or more pictures of YOU AND YOUR AROBOT. Make sure to include a view of the paint job and any accessories you have made.

■ Write a short (or long) paragraph explaining your robot's tasks, programs, accessories, and future expansion plans. Also include who you are and where you live.

■ Allow us to put this info on our web site and make you famous.

Send to the address below or use email.

SHIPPING ADDRESS	CONTACT US
Arrick Robotics 2107 W. Euless Blvd. Euless, TX 76040	Ph: (817) 571-4528 Fax: (817) 571-2317 info@robotics.com

Return Policy

Customer:

Please look over your ARobot very carefully. If you decide that it's beyond your skill level or not exactly what you expected, please return it promptly without continuing.

** Note **
If you begin the assembly process we can not make a refund.

When returning merchandise, please include a copy of your packing list and your contact information such as phone number or email address. Shipping charges are not refunded.

www.robotics.com

Chapter 8

From Whiskers to Wheels

In This Chapter

▶ Installing the whiskers

▶ Attaching the drive motor to the robot body

▶ Wheeling and steering

*I*f you're the kind of person who actually looks forward to assembling the new stereo or bicycle on Christmas morning, or if figuring out how to put part A into slot B makes your heart sing, this is the chapter for you. (I assume you are such a person or robot building probably wouldn't appeal to you.)

After prepping ARobot, it's time to get into assembling the parts. In this chapter, you give your robot whiskers so it can sense its environment, add a drive motor to the chassis to power your electronic friend, add wheels to set ARobot in motion, and install a steering system to imbue it with a sense of direction.

Starting with Whiskers

After you've prepared the various parts of ARobot, the first step in assembling them is to mount the two whisker wires. *Whiskers* are hair-like, flexible wires used by ARobot to detect walls and other solid objects. In the whisker assembly process, you'll bend and mount the whiskers, and then test them for continuity.

To mount the whiskers, follow these steps:

1. **Locate the four mounting holes near the front center of the robot body (two small ones and two large ones).**

    The front two holes are slightly bigger so that you can insert plastic shoulder washers.

2. **Scratch off any paint around the smaller holes.**

3. **Bend the whisker wires, as shown in Figure 8-1.**

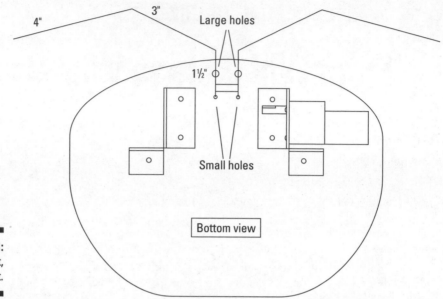

Figure 8-1:
Bend it,
shape it.

4. **Mount the two whisker spacers, according to the illustration in Figure 8-2.**

   Place the ground wire from the body cable under one of the whisker spacers and a #4 star washer under the other.

5. **Mount the whisker brackets, which are connected to the body cable.**

   The bracket connected to the white wire should be mounted to the robot's right side; the black wire to the left side.

6. **Pay special attention to the plastic insulating washers; these brackets must not make electrical connection to the body.**

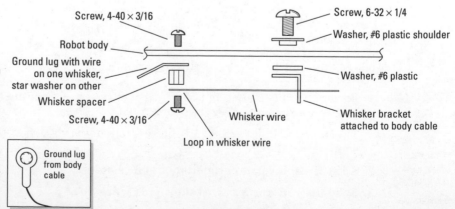

Figure 8-2:
Connect
whisker
wires with
the screws,
washers,
and spacers
provided.

7. Slide the whisker wires through the bracket holes and mount the wire's loop to the whisker spacer using a 4-40 x 3/16 screw.

8. Bend the whisker so that it rests in the center of the bracket holes.

9. If you have a multimeter, test for continuity between the whisker and body. Then test for no continuity between the whisker bracket and body.

When you've finished assembling the whiskers, place a small piece of tape around the ends to prevent the wires from poking people or unsuspecting pets that they run into.

Attaching the Drive Motor

The drive motor, also called a gear motor, attaches to the robot body by means of two motor brackets. The motor is the mechanical heart and lungs of the robot which, once provided with power from the batteries, runs the little guy.

The drive motor mounting process is simple, but you might have to bend some of the hardware to get the mounting holes to align.

Follow these steps to mount the drive motor:

1. Locate the two motor brackets; there will be a left side bracket and a right side bracket.

2. Mount the drive motor to the left bracket by screwing three 4-40 x 3/16 screws into the bracket.

See Figure 8-3 for guidance.

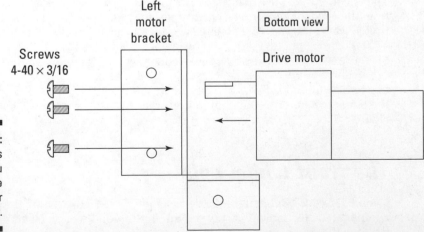

Figure 8-3: This is how you mount the left motor bracket.

3. **Attach both brackets to the robot body using the six motor-bracket mounting holes.**

   Use three 4-40 x 3/16 screws, three #4 star washers, and three 4-40 nuts to mount each motor bracket.

Dealing with Wheels

The next logical things to assemble are the wheels and other paraphernalia that make the wheels go. These include:

- **Encoder sensor:** This is the device that the motor uses to sense position and speed. It consists of a wheel with slots that are detected by an optical sensor.

- **Front wheel axle:** Just like the axles on your car, the front wheel axle is where the front wheels are mounted, and it is used to transfer power from the drive motor to the wheel.

- **Rear steering alignment:** ARobot has rear wheel steering. At the rear of the robot chassis you have two wheels on axles, plus a steering motor that — you guessed it — steers your robot. The steering system consists of the motor, linkages that go from the motor to the steering arms, and a steering horn that connects the servo motor to the linkages.

The encoder sensor

The front wheel optical encoder system tells the robot how far it's traveled, which is necessary for navigation. This simple system works like this: When the drive motor turns the axle, the drive wheel and the encoder wheel both rotate. The encoder wheel has many small teeth, which are detected by a light beam from the encoder sensor. The encoder sensor transmits the signal to the controller, where the teeth are counted and the distance traveled is determined.

In Figure 8-4, you can see how the encoder sensor assembly should look.

Use two 4-40 x 3/8 screws, two #4 star washers, and two 4-40 nuts to attach the encoder sensor to the right motor bracket, as shown in the illustration in Figure 8-5.

The front wheel axle

I won't kid you: ARobot's front axle assembly (see Figure 8-6) is the most complicated mechanical part of the robot. But don't worry: I'll guide you through the process.

Figure 8-4:
The encoder sensor assembly.

For this system to work correctly, the encoder wheel must rotate inside the encoder sensor arms while the robot's drive motor is turning. Note that the encoder wheel must be fully inside the two arms of the sensor, but it must not touch the sensor at all. To achieve this alignment, you might have to gently bend the bracket (refer to Figure 8-4) on which the encoder sensor is mounted towards the encoder wheel. ***Remember:*** Make sure that the wheel doesn't touch the sensor when you're finished.

Right motor bracket

Screw,
4-40 × 3/8

Figure 8-5:
Make sure the dots on the encoder match the placement on this figure.

Encoder sensor   Dots

Star washer, #4   Nut, 4-40

Figure 8-6:
Front
wheel axle
assembly.

If the encoder wheel does touch the sides of the sensor and bending the bracket doesn't do the trick, loosen the nuts holding the encoder wheel, center the wheel, and then retighten the nuts.

Be careful not to damage the threads on the axle or it will become useless to any robot on the planet.

Carefully follow these instructions for assembling the front wheel components:

1. **Locate the wheel with the threaded bore and red mark on the hub.**

2. **Thread the front axle through the wheel so that it protrudes ½ inch on one side.**

3. **On the short axle side, slide on a star washer, screw on the coupling, and tighten it.**

4. **On the long axle side, slide on a star washer, screw on a nut, and tighten it.**

5. **On the long axle side, screw on a nut, a washer, and then another nut.**

6. **Slip the bronze bearing onto the long side of the axle, flange (wide end) first.**

7. **Insert the long axle side through the hole in the right motor bracket, and attach the coupling onto the drive motor shaft.**

   The bearing should fit in the hole. If it doesn't, you may have to use a knife to enlarge the hole.

8. **Move the nut-washer-nut combination you installed earlier so that they push against the bronze bearing.**

9. **Install the encoder wheel with nuts and washers, as shown in Figure 8-7.**

   The encoder wheel should ride in the slot in the encoder sensor but not touch it.

10. **Tighten the set screw in the coupling against the motor shaft.**

11. **Tighten the wheel against the coupling.**

12. **Tighten the other nut against the wheel, and then tighten the two nuts together.**

13. **Tighten the nuts against the encoder wheel.**

Make sure that the components are fitted tightly to prevent loosening while the robot is driving around your backyard.

If the drive wheel or collar isn't tight on the front axle, the axle might gradually shift over time. Make sure that all the nuts are tight and that the whole front wheel assembly has the correct hardware placement.

Rear steering alignment

Imagine driving along down the highway in your Chevy or Ford and suddenly discovering that you have no steering wheel. This would be a bad thing. ARobot also has to have a way to steer itself around, so it comes with a unique rear-wheel steering system. In this system, a single RC servo motor is linked to control arms on each rear axle. Make sure that the linkage wires don't bump into the steering servo motor.

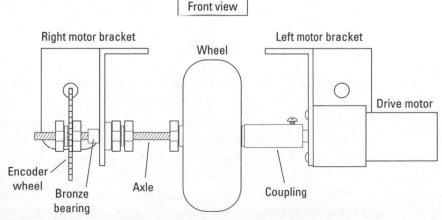

Figure 8-7: A star washer and 10-32 nut secure the encoder wheel to the right motor bracket.

To assemble this steering system and the rear wheels of the robot, follow these steps:

1. **Bend both 4-inch wires into steering linkages with pliers, using the illustration in Figure 8-8 to guide you.**

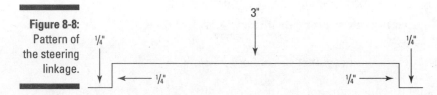

Figure 8-8: Pattern of the steering linkage.

2. **Attach the horn to the steering motor using a single screw.**

   Turn the horn in both directions and determine the center position. In the center position, the horn should be oriented as shown in Figure 8-9.

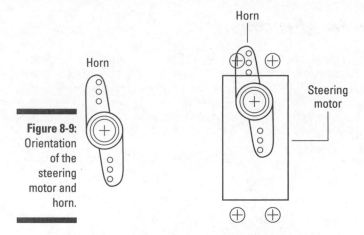

Figure 8-9: Orientation of the steering motor and horn.

3. **Mount the steering motor to the robot body using four screws, washers, and nuts, as shown in Figure 8-10.**

4. **Mount the rear axles to the robot body using collars with 6-32 x 1/4 screws inserted in them, as shown in Figure 8-11.**

5. **Mount the wheels to the right and left axles using collars with 6-32 x 1/4 screws (refer to Figure 8-11).**

6. **Attach the linkages to the steering motor and then to both steering arms, as shown in Figure 8-12.**

7. **Attach the steering arms to the axles.**

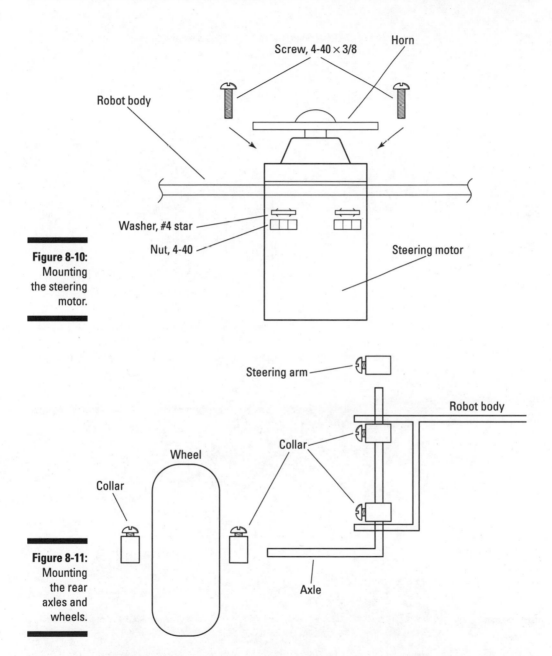

Screw, 4-40 × 3/8

Horn

Robot body

Washer, #4 star

Nut, 4-40

Steering motor

Figure 8-10:
Mounting
the steering
motor.

Steering arm

Robot body

Collar

Wheel

Collar

Collar

Figure 8-11:
Mounting
the rear
axles and
wheels.

Axle

Now you need to align the steering system (see Figure 8-13). The output shaft of the servo motor has a plastic piece called a servo horn that you attached the linkages to. Turn the horn by hand and notice the extent of its movement — it should be about 120 degrees. With the horn in the approximate center of its movement, adjust the arm on each axle so that the wheels are pointing forward.

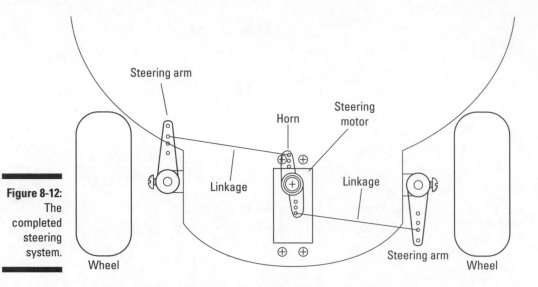

Figure 8-12:
The
completed
steering
system.

Now that you have the guts of the beast assembled, move on to the brains and power source for ARobot in Chapter 9. There, you install an onboard computer called a controller and add batteries to give your mechanical buddy a charge.

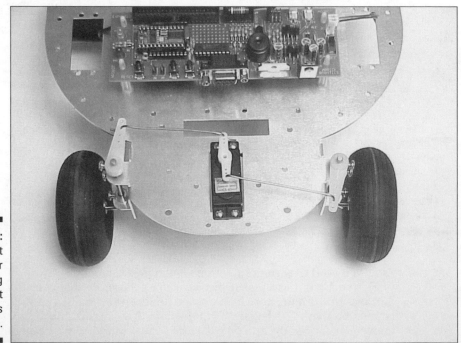

Figure 8-13:
Check that
your rear
steering
alignment
looks
like this.

Chapter 9

If I Only Had a Brain

The whole point of building robots is to make them more intelligent than your average boulder. To do that, programmable robots need something called a controller, so you can download software programs to the robot and get it to do things.

After you have the body and brains of the robot in place, they won't do a thing without power, so the last part of this chapter deals with batteries and other power sources that, like Frankenstein's bolt of lightening, will finally bring your robot to life.

Adding the Controller

Computers are everywhere — in our appliances, in our cars, and on our desktops. These computers differ in memory capacity and processing power. ARobot uses a small computer from Parallax called the Basic Stamp 2 (BS2) controller, a popular item among hobbyists who build smart devices. You can program the BS2 from your desktop computer to allow ARobot to perform automatically and make its own decisions.

Like any computer — big or small — the BS2 includes memory to hold the program and data along with a processor to execute the instructions. Also included is a set of input/output (I/O) pins that control devices such as motors, lights, and speakers.

ARobot's controller board is designed to accept the Basic Stamp 2 controller, as shown in Figure 9-1. (For more information on the BS2, go to the manufacturer's Web site at `www.parallax.com`.)

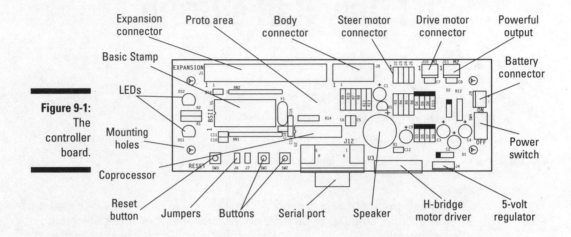

Figure 9-1:
The
controller
board.

You write programs for ARobot on your desktop computer using an easy-to-learn language called PBasic. You'll find that PBasic provides all the instructions you'll need to make ARobot do its stuff.

After you write a program, you download it to the robot's BS2 controller for execution. Downloading is accomplished using a simple serial communications cable, which you can remove during the robot's operation.

ARobot's user guide will not teach you programming concepts or details about how the BS2 works; you can find many of those topics covered in Chapter 6. Also, many programming books are available that explain the details of the PBasic language and how to use the BS2.

Selecting just the right controller

I'm all for trying whatever is new and exciting, but sometimes the tried-and-true option is the best way to go. That's the case with robot controllers. The BS2 has been around for several years and is popular. Many mail-order companies and electronic shops carry the BS2 and similar controllers. That track record, along with the modularity that the BS2 offers, makes it a good option for control projects — and perfect for small robots.

Because of the popularity of BS2, several companies offer BS2-compatible controllers. Although some of these are plug-in replacements, others have minor differences, such as the programming language used, the amount of electrical current they draw, and their memory capacity. Parallax and other companies have BS2-compatible parts that offer various improvements in performance and capacity. Table 9-1 offers a comparison.

Table 9-1	Controller Comparison Chart			
Feature	*BS2*	*BS2sx*	*BS2p*	*BasicATOM*
Speed (instructions/second)	4K	10K	12K	33K
Memory	32 bytes	32 bytes	38 bytes	368 bytes
Program size	2K	16K	16K	14K
Current consumption	8ma	60ma	40ma	5ma

Parallax makes all these products with the exception of the BasicATOM, which is made by Basic Micro (www.basicmicro.com). As you can see, compromises have been made in speed, memory, and current consumption to get certain features. In addition, differences between the models often require program changes.

The program code and circuits in this book are designed and tested to work specifically with the BS2, so I suggest you use that controller unless you're willing to make the changes that might be required to make another model work correctly.

Installing the controller board

The controller board is command central for your robot. It's where you install the computer chip called the BS2 controller, and it's also where you will find the power connector, serial port connector, and expansion board connector.

Before you add the brainy little controller chip to the controller board, you have to mount the controller board on the robot body, following these steps:

1. **Mount four plastic spacers on the four mounting holes on the robot body using 6-32 x 1/4 screws.**

   You'll see two sets of these holes; use the ones toward the rear of the body.

2. **Snap the controller board in place onto the spacers.**

Installing the controller

Unlike programming your VCR, installing a BS2 chip in ARobot's controller board isn't hard. It just involves placing a chip in a socket and matching up the pins correctly. Before you begin the procedure, however, you should read these precautions to ensure that things go smoothly:

- ✔ **Turn the power off:** Never install a BS2 — or any chip for that matter — with the power turned on. Doing so could ruin a perfectly good part . . . and your day. Also, never remove the BS2 with the power turned on (for the same reason).

- ✔ **Avoid bent pins:** Pins were made to be straight, so make sure that all the pins on the BS2 are in an unbent state. If they're not, they might bend further (or even break) while you're inserting the chip into the socket.

- ✔ **Match your pins correctly:** Anyone can easily put the wrong pin in the right place, so watch your alignment carefully. Pin 1 on the BS2 must match pin 1 on the socket. On ARobot's controller board, notice the 1 that identifies pin 1 on the socket. Then on the BS2, notice the small half-circle indicator at the top of the chip. Finally, put them together.

Now it's time to install the BS2:

1. **Remove the battery cable from the controller.**

2. **Use Figure 9-2 to locate the socket on the controller board that matches pin 1 on the BS2.**

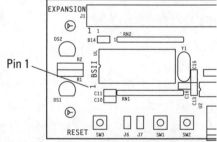

Figure 9-2:
The pin indicator helps you guide the pin into the socket.

3. **Insert the pins into the sockets, making sure that no pins are bent (see Figure 9-3).**

Connecting the battery cable

Although you don't install batteries yet (that happens in the "Turning ARobot On" section, later in this chapter), it's time to connect the battery cable to the controller board. The battery cable has a battery snap on one end and a 2-pin MTA connector on the other. Pin 1 has a red wire providing +12 volts; pin 2 has a black wire that's the ground wire (that is, 0 volts).

Follow these steps to hook up the battery cable:

1. **Plug the 2-pin MTA connector onto the controller board's power connector.**

2. **Route the cable through any nearby hole in the body.**

3. **Place one side of the provided Velcro strip on the side of the battery holder that provides a long, smooth plastic area.**

4. **Place the other side of the Velcro strip on the bottom of the robot, anywhere where the cable can reach it.**

5. **Connect the cable to the battery holder using the battery snap.**

6. **Mount the battery holder to the bottom of the robot.**

You'll feel the desire to install batteries before you finish building a robot just to see if there's any life to it, but resist! If you do, the wires will have voltage on them and you could create a short, causing the batteries to drain. Install the batteries only when the instructions call for them.

Figure 9-3:
Install
the BS2
carefully.

Correct connections with MTA

MTA connectors are small connectors that are popular for making connections from a PC board to a cable. They usually have between 2 and 20 pins and are simple to insert and remove. MTA connectors are *keyed,* which means they're designed to be inserted in one direction only. ARobot uses these types of connectors for the battery cable, encoder sensor, and the drive motor cable.

Connecting the body cable

The body cable connects the controller board to the encoder sensor, whiskers, and ground lug. It's already assembled and tested, so all you have to do is to make the connections. Do so using this simple process:

1. **Route the body cable up through the large rectangular opening in the robot body near the controller board.**

2. **Plug the cable into the body connector on the controller board.**

3. **Plug the MTA connectors onto the encoder sensor pins by matching the dots shown in Figure 9-4.**

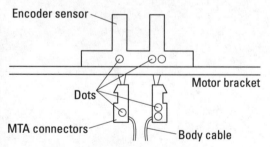

Figure 9-4:
Connecting the encoder sensor with help from the dots.

Encoder sensor

Motor bracket

Dots

MTA connectors

Body cable

Finishing Touches

A few simple steps are left to tidy everything up and get ARobot ready to power up and accept programs that tell it what to do.

Perform these three simple procedures:

1. **Route the steering motor cable up through the large rectangular hole in the robot body near the controller board.**

2. **Connect the steering motor cable to the 3-pin connector on the controller board (near the body connector). Align the white (or yellow) wire to pin 1.**

3. **Attach the body cable to the robot using plastic wire ties.**

   You can use any unused holes in the body to do this.

4. **Route the drive motor cable up through the same opening in the body.**

5. **Attach the drive motor cable to the motor connector on the controller board.**

   It only plugs in one way — the right way!

That's it. ARobot is all assembled. Now it's time to power it up.

Turning ARobot On

You've endured a multitude of fancy parts and metal screws, and now you have a nice little robot to show for it. Time to bring your robot to life.

ARobot comes with a plastic battery holder that carries eight AA batteries; this configuration produces approximately 12 volts when using alkaline batteries and about 9.6 volts when using rechargeable batteries. You install the batteries by simply popping them into the battery holder with the plus and minus signs facing as indicated on the holder. Of course, don't forget to charge rechargeable batteries before installing them.

Batteries and other power sources affect not only your robot's performance but also your bank account. Turn to Chapter 3 for details on several battery options.

Giving power to ARobot involves a flip of the power switch, which is located next to the power connector on the controller board. Following the logic of street lights, the on position sports a green dot, and the off position has a red dot.

Be sure you turn the power off when you're attaching cables, installing batteries, or installing parts.

When you turn the power on to ARobot, the program in the BS2 controller begins to run automatically. At any time, you can press the reset button on the controller board to restart the program from the beginning. Often, programs are written to wait until one of the buttons is pressed to begin operation. This prevents the robot from taking off right as the power switch is turned on.

Now that you've brought your robot to life, it's just dying to have you download some programs into it so it can get busy. Conveniently, that's the topic of Chapter 10.

Chapter 10

Introducing Your Robot to Your Computer

*T*he *programmable* part of a programmable robot relates to its capability to connect to your computer and lets you download programs into it. These programs can control the movements of the robot as well as cause the robot to gather data and make decisions based on that data. For example, with the correct sensor attached, the robot could sense the light level, and a program could make it move towards or away from the light source.

To write programs for ARobot, you use the Basic Stamp editor. You use the editor also to download the finished program to ARobot's controller board for execution. In addition, the editor offers some convenient debug features.

Working with the Basic Stamp Editor

Right about now, you're probably getting impatient to see some robot action resulting from all your hard work. After all, you've built ARobot, installed the BS2 controller, and selected batteries. You're getting close to your payoff: It's time to apply the juice and begin to build the brain of the beast.

You first power up the robot by using the slide switch next to the battery connector. When you do, don't be alarmed when nothing happens . . . that's okay. Really. Remember that the program memory in the BS2 is probably empty. Before you can get anything out of this little fella, you have to connect ARobot to your computer and download a program into your robot's memory.

Installing the software

Before you can download a program, you need a *downloader program* (see how the software industry has us all in their power?). Parallax (`www.parallax.com`) provides an editor/downloader program called Basic Stamp Editor that provides an environment to create and debug programs as well as download them to your robot. This editor works under most versions of Microsoft Windows as well as a few other operating systems.

To install the Basic Stamp editor program, simply follow the instructions on the CD or click the EXE file after downloading.

If you're a Macintosh user (and are wondering whether this is a PC-only deal), I have good news for you: A Basic Stamp editor for Macs is also available from Parallax. How about Linux, you ask? Parallax has made a *tokenizer* available for Linux, and you should be able to eventually get an editor/downloader for that operating system as well.

Opening the editor

After the editor is installed, opening it is as simple as clicking its icon. A window appears with an editing space and a menu bar across the top, as shown in Figure 10-1.

Loading or creating a program

You can create a program by manually typing in the editor window, just as you would in your favorite word processor, but it's much simpler to begin with an existing program. Starting with a previously tested program such as those supplied on the ARobot floppy also makes it easier to solve problems because you can eliminate the program code as a possible suspect.

To begin with an existing program, choose File➪Open. Then click to select the desired file. (Programs you can run with the Basic Stamp 2 editor have the BS2 file extension.) In Figure 10-2, I loaded the wander.bs2 program.

Downloading a program to ARobot

After you load a program into the editor, take a look at the code. Near the beginning of most programs are comments describing the program's function and possibly providing operator instructions.

Figure 10-1:
The Basic
Stamp
editor
used to
create and
download
the robot
programs.

To download the program to ARobot, make sure that the BS2 controller chip is properly installed, and then follow these steps:

1. **Turn on power to the robot.**

2. **Choose Run⇨Run or click the blue right-facing triangle icon under the menu bar. (Keyboard lovers can download the program by pressing Ctrl+R.)**

    • **If the download is successful:** The program can communicate with the robot and a small dialog box appears to monitor the downloading progress, as shown in Figure 10-3. The process takes only a few seconds.

    • **If the download is not successful:** A message appears describing the situation. In Figure 10-4, the trouble is probably related to communications. The next section describes typical problems.

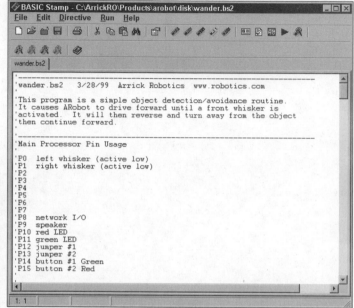

Figure 10-2:
The
wander.bs2
program
has been
loaded into
the editor.

After the download is complete, the robot immediately begins executing the program. Depending on the program, you may have to start the operation by pressing a switch. Or, the robot may just take off, so have your robot leash handy. If you have your robot on a table, this could result in a visit to the robot ER.

Figure 10-3:
The pro-
gram is
downloaded
in the blink
of an eye.

The first thing you'll probably want to do after downloading is remove the communications cable. In most robot programs, the robot moves around so you won't want a cable hanging off the back. If the robot begins to move immediately, pick it up and remove the communications cable, and then set it back down.

Figure 10-4: Down-loading errors are easy to correct.

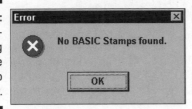

At any time, you can turn the power switch off and the robot will shut down. Don't worry, the program won't be lost. Also at any time, you can press the reset button on ARobot's controller board to restart the program from the beginning. You'll find this reset button useful when debugging your first programs. (Both the reset button and the power switch are labeled in the first figure in Chapter 9.)

Troubleshooting downloading

I'll tell you right now: If you can't figure out how to download stuff to your robot, it becomes a pricey little doorstop. Without downloaded programs, your robot can't do the simplest thing, so take some time to make sure that you're downloading successfully.

Downloading problems are usually an issue with the communication between your PC and the robot's controller or are something simple to solve, such as low batteries.

Here's the checklist for downloading troubles:

- ✔ **Check the cables:** Make sure the cable is connected correctly on both ends. In other words, are the things plugged in correctly?
- ✔ **Check the power:** Make sure that power to the robot is turned on. Remember, no power, no download.
- ✔ **Check for low batteries:** Some batteries have pressure indicators; if yours doesn't, use a voltmeter or a battery tester.
- ✔ **Check the battery pack:** Make sure the battery pack is plugged into the robot properly.
- ✔ **Check port settings:** Experiment with communication port settings on your computer. Use the editor's preferences menu (in the Edit menu) to experiment with com port settings.

Running the Built-In Programs

ARobot comes with three programs that allow you to test and play with your robot. The first program to use is the test.bs2 program, which allows you to test most functions and devices on ARobot, including the speaker, whiskers, steering motor, drive motor, encoder sensor, LEDs, switches, jumpers, and communications port.

The second program of interest is the straight.bs2 program, which helps you fine-tune the steering system by making the robot move forward and then backward at the push of a button.

The final program is the wander.bs2 program, which makes use of ARobot's whiskers for navigation. The robot will move around and back away from objects it encounters. The wander program is a good example of how ARobot can operate in the real world and it also provides a great starting point for more complex programs.

Testing the robot's functions

The testing program, test.bs2, can test most functions of the robot. Here's how you download this program to your robot:

1. **Make sure that the serial cable is securely connected to your computer and ARobot.**

2. **In the Basic Stamp editor, choose File⇨Open, and then choose the test.bs2.**

    You see the PBasic code that makes up the test program, as shown in Figure 10-5.

3. **Verify that the power to the robot is on.**

4. **Choose Run⇨Run or click the blue right-facing triangle icon under the menu bar.**

    A small window appears, showing you the downloading progress.

    • **If the download is successful:** When the download is complete, you'll hear several beeps from the robot — it's first sign of life!

    A new window — the Debug Terminal window — appears in the editor program at this point, as shown in Figure 10-6. Any data that's sent from the BS2's serial port appears as text in this window. This window is where the program displays data to help you in debugging programs.

- **If the download is not successful:** You get a message telling you what's wrong. See the "Troubleshooting downloading" section previously in this chapter to help you solve any problems that you encounter at this point.

Now that you've downloaded the test program, you can test most functions of the robot, including the LEDs, speaker, motor controller, pushbuttons, whiskers, and serial port. (If you need help finding the switches and jumpers, flip back to Figure 8-1.)

ARobot has two small jumpers, J6 and J7, located near the pushbutton switches. Each jumper consists of two prongs that are activated by using small shorting plugs that connect the prongs together. The BS2 controller can read these jumpers, just like the pushbuttons, to see whether they're on.

Figure 10-5: The test.bs2 program is downloaded to the editor.

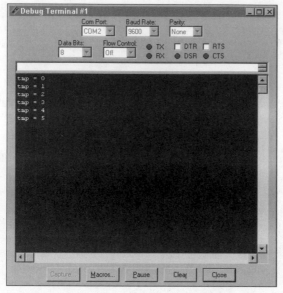

Figure 10-6:
The Debug
Terminal
window of
the Basic
Stamp
editor.

Follow these steps:

1. **Press switch 1 (SW1).**

   The green LED turns on and the steering (and all RC servo motors) moves left and right.

2. **Press switch 2 (SW2).**

   Both LEDs flash, the speaker beeps, and data appears on your computer's debug window.

3. **Install jumper J6.**

   Motor 2 (if you added one) moves forward. If you haven't added a device to the motor 2 port, ignore this test.

4. **Install jumper J7.**

   The green LED turns on and motor 1 (ARobot's front drive motor) propels the robot forward. The red LED also turns on and off every 128 encoder counts (roughly every 15 seconds).

5. **Press the left whisker.**

   This turns on the green LED and makes the robot beep once.

6. **Press the right whisker.**

   This turns on the red LED and makes the robot beep twice.

Use common sense to troubleshoot problems. For example, if the whiskers are wired backwards, the left whisker will do what the right whisker should be doing. If the steering motor or drive motor doesn't move, check the connectors. If the red LED doesn't turn on and off when the front motor is turning, check the encoder sensor connectors. If the test program starts out by moving the robot forward, make sure that you removed the jumper shorting plugs.

Adjusting the steering

After you verify that the various functions controlled by the test program work, get your ARobot on the right track by straightening the rear wheel steering system. You do this to determine whether your robot can move in a straight line, which is an essential robot skill.

A program called straight.bs2 makes this task simpler. When you download and start this program, it moves the steering motor to the center position and then drives the robot forward 10 feet and then backward 10 feet. You can tell by the motion whether the robot is driving straight or not.

The wheels are straightened by adjusting the steering arms attached to each rear axle. The arms are connected to the servo motor with wire linkages. Follow these steps to load and run the straight.bs2 program:

1. **Load the straight.bs2 program into the Basic Stamp editor.**

   To do so, choose File➪Open, and then choose the straight.bs2 program. You see the PBasic code that makes up the straight program.

2. **Verify that the power to the robot is on.**

3. **Choose Run➪Run or click the blue right-facing triangle icon under the menu bar.**

   The program doesn't begin running automatically. Instead, it waits for you to perform the next step.

4. **Press SW1 on the controller board to start the motion.**

   The steering motor moves to the center position, and the robot moves forward and then backward.

5. **If the robot doesn't move in a straight line, adjust the steering arms and press SW1 again.**

   Test for straight motion as many times as needed.

Wandering around with the Wander program

The wander.bs2 program uses ARobot's mighty brain, along with the drive motor and whiskers, to navigate around. This is an excellent program for understanding how ARobot operates.

1. **Load the wander.bs2 program into the Basic Stamp Editor.**

   To do so, choose File⇨Open, and then choose the wander.bs2 program.

2. **Verify that the power to the robot is on.**

3. **Download the program to the robot by choosing Run⇨Run (or clicking the blue right-facing triangle icon under the menu bar).**

   Rather than start automatically, the wander program waits patiently. This small detail prevents the robot from jumping off a table or down the stairs.

4. **Press SW1 to start the robot.**

   The robot begins to move forward while reading the whiskers. When the robot encounters an object, such as a table leg or wall, it moves backward away from the object and then begins to move forward again.

If the left whisker is activated, the robot backs away to the right. If the right whisker is activated, the robot backs away to the left. This allows ARobot to move away from objects, although it may require several attempts.

When writing a new program, you might want to start with the wander program because it incorporates many program pieces that will help you understand how to make the robot do what you want. Building your own version of the program is as simple as cutting and pasting various pieces, and then adding your own PBasic commands as needed. This is by far the fastest and most enjoyable way to learn programming and to get your robot to be all it can be.

Troubleshooting

Even though the ARobot kit makes building a robot about as easy as it can be, you still might encounter pitfalls and moments when you'd like to throw your robot in the disposal, piece by piece. Don't.

In this section, I tell you how to solve a few common problems . . . and thus save your robot's life.

Murphy's Law is definitely in force in the world of robotics, and now and then if it can go wrong, it will. In this section, I present some common problems the first-time robot builder can run into.

Tricky whiskers

In my house, I can trace many strange phenomena to my cat: the paw prints on my car hood, the shampoo bottle spilled all over the bathroom counter, and on and on. With robots, many strange problems can be traced back to non-working (and non-feline) whiskers, especially if you're encountering navigation problems.

If your robot doesn't move or constantly goes in reverse, one or both of the whiskers might be stuck in the on position.

Here are some common whisker problems:

✔ **Bent whiskers:** If the whisker has been bent by accident, it may be constantly activated and result in strange robot behavior. Make sure that the whisker wire hangs in the center of the bracket hole and doesn't touch the bracket until the whisker encounters an object.

✔ **Crossed whisker wires:** If the left and right whisker signals are crossed, some bizarre navigational behavior can occur. Check the whisker wiring by using the test.bs2 program.

✔ **Uninsulated whisker brackets:** Each whisker bracket must be insulated from the robot's metal base; otherwise, the whisker will be on all the time. What problem this causes depends on the software involved. Refer to the drawings in ARobot's user guide to make sure that you correctly installed the insulating washers.

Steering gone astray

Is your robot weaving around the room like an inebriated driver who just failed a Breathalyzer test? Robots don't generally drink, so it could be a steering problem. Steering problems are pretty easy to solve and usually relate to the mechanical connections of the motor hardware or the electrical connection of the motor itself.

If you have problems that you think are related to steering, check out this list.

✔ **Reversed servo horn:** The linkage between the steering servo motor and the steering arms on each rear axle could be installed incorrectly, which could cause the steering to be reversed. Check that the linkages, arms, and horn are connected as shown in ARobot's user guide.

✔ **Non-functioning steering motor:** If the steering servo motor doesn't move at all when a program commands it to, check out these possibilities:

• The motor should be connected to connector J2.

- You might have reversed the servo motor connector, so try the other direction.

- Check the software and verify that the code is correct. It takes only a single typo to prevent the motor from working.

Getting the drive motor right

ARobot's front drive system is the most complex mechanical part of the robot. You have to get many pieces of hardware and wiring just right. (And that's not all — you also have to download software to run the thing.)

If you're having problems that you suspect are linked with the drive motor, check these items:

- ✔ **Wheel hardware:** The wheel hardware, including all the nuts, must be tight (or the nuts will spin freely). Loose hardware sometimes results in free spinning in only one direction and proper robot movement in the other.

- ✔ **Encoder sensor cables:** The encoder sensor cables must be installed correctly. Check the drawings in ARobot's user guide and match up the dots with the connectors. Without a properly working encoder, the robot will usually move without stopping because the distance traveled can't be measured.

- ✔ **Encoder wheel:** The encoder wheel must be centered between the sensor arms but never touch them. The encoder wheel must be fully between the sensor arms; if necessary, bend the bracket that holds the sensor toward the wheel to correct this.

- ✔ **Axle collar:** The collar that connects the axle to the drive motor might have a loose screw. Tighten it so that the front wheel will turn when the motor is on.

Controller, this is the tower. We have a problem . . .

Just think what would happen if your brain stopped suddenly. (Well, mine does now and then, but you're probably unfamiliar with the phenomenon.) The *controller* is the brain of this robot beast, and you have to take good care of it.

Several things can cause a nonfunctioning controller, which results in a brain-dead robot. Check the items on this list if your controller doesn't seem quite up to snuff:

- ✔ **Check your program:** Before you conclude that you have a hardware problem, double-check your program. Better yet, use one of the pretested programs from this book or from the Web site.

- ✔ **Batteries:** A single bad battery can ruin the entire batch. If you have to, try a new set.

- ✔ **Connections:** The battery pack cable must be plugged securely into the controller board and battery pack.

- ✔ **BS2 installation:** You have to have installed the BS2 correctly. Check the orientation against the drawings and pictures in ARobot's user guide or read through the "Adding the Controller" section in Chapter 9.

- ✔ **Bent pins:** You could have a bent pin on the BS2. If so, unbend it carefully Breaking a pin on the BS2 is an expensive mistake.

- ✔ **Power:** Sorry, I know I risk embarrassing you, but I have to ask: Have you turned on the power?

The dead robot

The easiest type of problem to solve with electronic or computer equipment is when it's completely dead. Lack of power is usually the culprit in those situations. It's important that batteries are installed correctly, so make sure they match up with the indicators on the battery holder and make sure they're fresh. A dead battery equals a dead robot.

Part IV
Augmenting Your Programmable Robot

The 5th Wave   By Rich Tennant

"Do you remember which military web site you downloaded your Bot software from?"

In this part . . .

Webster's Dictionary defines *augment* as "to make or become greater." That's what the chapters in this part are all about: making your robot greater by adding functionality to it. In other words, you'll take that mechanical guy and make it do nifty things.

Here's where you discover how to use programs and add-on gadgets that allow your robot to sense light, temperature, and motion. You'll enable your robot to talk to you and even "look" at its surroundings with an on-board camera. Finally, you play around with adding remote control to your robot buddy.

Let the augmenting begin!

Chapter 11

Expanding Your Robot's Universe

Anybody who has had a Barbie or GI Joe doll knows that the way they come out of the box is pretty darn dull. The whole point of these things is to buy them outfits or armored divisions to play with. In fact, the cost of the original doll is nothing compared to the cost of the accessories you buy for them.

Your ARobot, though a neat little guy in his own right, practically begs to be expanded. That's why ARobot is designed with extra controller ports, predrilled holes, and special connectors to allow you to add many sensors and actuators such as a moveable head or a gripper.

In this chapter, I show you how to plan for ARobot's expanded wardrobe by getting all the pieces in place to accommodate very cool accessories.

Adding Rear Whiskers

You don't add whiskers to a robot to make it look like your pet cat. You add whiskers to give it a sense of touch. A robot without a sense of touch would go slamming into things right and left and do itself and your house no good at all.

Whiskers are mechanical switches that simply provide the robot with an on/off status. On can mean the whisker is in contact with an object such as a table leg. Off can mean the whisker is in the clear, for example, and the robot can react accordingly (well, actually, it can react any way you program it to react). Whiskers are easy to read with a program and are an invaluable aid for navigation.

ARobot comes supplied with two independent front whiskers, as shown in Figure 11-1. This section explains how to add rear whiskers to improve navigation when your robot decides to go into reverse.

Figure 11-1: Front whiskers waiting for action.

All about whiskers

So what does a whisker look like? The whisker itself consists of a very thin metal wire. The whisker wire is mounted to the robot's base plate, which is attached to the ground terminal of the battery through the body cable. The whisker is positioned through a hole in a small metal bracket, which is connected to a wire that goes to the robot's controller board.

The metal bracket must be insulated from the robot's base plate. Otherwise, ARobot's program will constantly read the whisker as activated.

Normally, the whisker wire doesn't touch the metal bracket. But when something comes in contact with the whisker and moves it enough, the wire does make contact with the bracket. Figure 11-2 shows a close-up of the whisker bracket, wire, and mounting.

Logic levels

In digital electronics, you often hear talk about logic levels. A digital signal can have two states: logic 0 (zero) and logic 1 (one), which are defined by voltage levels. A logic 0 is close to ground and a logic 1 is close to +5 volts. These states can be read with a voltmeter as long as they don't change too quickly. When the Basic Stamp 2 controller reads an input pin, it reads either a 1 or a 0, indicating the logic level present at the pin in question.

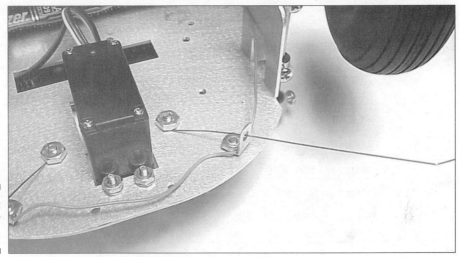

Figure 11-2:
Whisker
nitty-gritty.

The bracket is connected to ground and can be read by the controller as a logic 0. You can write the robot's program to move away from an object when logic 0 is read.

Collecting the parts

Don't go pulling the whiskers off your cat for this project — they won't work for our purposes and your cat will definitely resent it. Whisker sensors can be constructed with simple parts that you can find at most electronics and hardware shops, as shown in Figure 11-3.

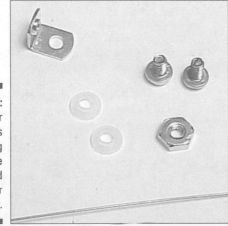

Figure 11-3:
Whisker
parts
waiting
to be
assimilated
into your
robot.

Table 11-1 lists the parts you need to complete two whisker sensors.

Table 11-1	Rear Whiskers Parts List		
Part	*Vendor*	*Quantity*	*Approximate Price Each*
Bracket, 534-614	Mouser	2	$0.30
Nut 6-32, 561-G632	Mouser	2	$0.06
Screw 6-32x1/4, 561-P632.25	Mouser	4	$0.07
Plastic washer, 561-06004	Mouser	4	$0.07
Steel whisker wire .010-.030 diameter	Hardware store	18"	$2.00

Installing the whiskers

You'll be adding two rear whiskers, but they'll be wired together to act as a single sensor. This saves a valuable port pin on the robot's controller and provides all the information necessary to detect objects while your robot is moving around the world in reverse.

Cooking from scratch

An enterprising robot builder can create a whisker sensor from scratch instead of buying off-the-shelf components from an electronics vendor, although it can be difficult. For the whisker wire, you can use any thin metal wire such as a metal guitar string or the kind of wire used to hang pictures. The wire must be thin and have the right amount of stiffness to work correctly. The machine screws and washers can be purchased at a hardware store.

The bracket is the most difficult item to find and may have to be fabricated out of thin metal stock. Your local home improvement store is likely to have bins full of small, unpainted, metal brackets made for door hardware or to hang pictures, and some of those could be used as whisker brackets.

If you're really ambitious and want to fabricate your own bracket from scratch, start with a ½-inch-by-½-inch piece of thin metal that's not painted. The metal must be thin enough to bend in the middle to make an L shape. Drill a small ⅛-inch hole in the center of both sections — one for the mounting screw and one for the whisker wire to go through.

Wherever you get it, it's important that the bracket be conductive, just like the whisker. Otherwise an electrical connection will not be made and the whisker won't work. Use an ohmmeter to check for conductivity.

An important detail to pay attention to when installing a whisker is the mounting of the bracket. The bracket must be insulated from the base plate to prevent the bracket from being grounded out on the robot's base. Use the insulating washers under the screw head as well as under the bracket to prevent this.

It's best to use a shoulder washer to insulate the screw from the inside of the hole. Another option is to use a plastic screw and one insulating washer under the bracket.

Here are the steps to install the whiskers:

1. **Solder a 4-inch wire between the two brackets, as shown in Figure 11-4.**

2. **Solder the wire from the body cable to one bracket.**

3. **Locate the mounting holes which will be near the steering servo motor at the rear of the base plate.**

4. **Install the bracket with insulating washers, as shown in Figure 11-5.**

5. **Make a loop in the whisker wire to go around the screw, as shown in Figure 11-6.**

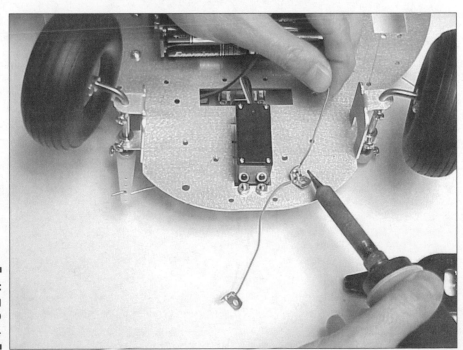

Figure 11-4: Soldering the wire to the bracket.

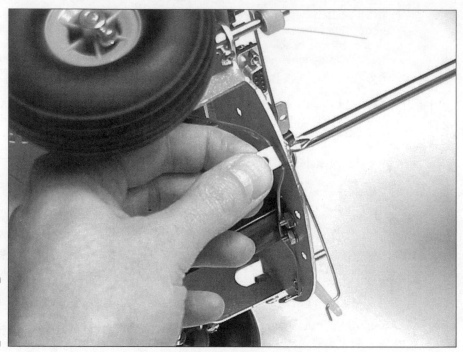

Figure 11-5:
Mounting
the bracket.

6. **Use the screws and spacer to mount the whisker wire to the robot's base.**

7. **Bend the whisker to aid object detection.**

   Usually a single bend of 45 degrees will allow the whisker to be activated from various angles.

8. **Bend the whisker so that it's not touching the bracket.**

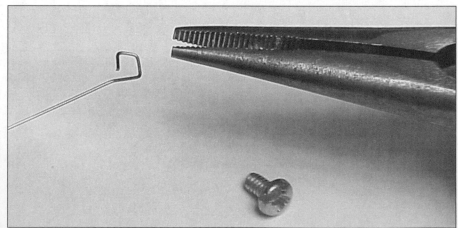

Figure 11-6:
Creating a
loop in the
whisker.

ARobot's body cable has an extra wire used to carry the signal from the whisker to the controller. To attach this cable, follow these steps:

1. **On the 10-pin body cable, locate pin 3, which is gray in color.**

2. **Peel the wire away from the others so that it can reach the rear whisker bracket mounting hole.**

3. **Strip the wire and solder it to the whisker bracket, as shown in Figure 11-7.**

Controlling the whiskers

To wake up your robot's whiskers and get them twitching (well, they won't actually twitch, but they will sense things), you have to download some code. Two programs are involved in making the rear whiskers do what you need them to do. The first tests the whiskers to make sure they're operating properly. The second uses the whiskers to sense an object and back up the robot when an object is encountered.

Waking up the rear whiskers

You can do a lot of things to make use of the rear whiskers, but before you write a large, ambitious program, start with a small test program that verifies that the whiskers are ready to sit up and take notice.

Figure 11-7:
The completed whiskers.

The rear whiskers are connected to P2 on the Basic Stamp 2 controller. Listing 11-1 is a short program that tests that the connection is good by lighting the LED when the whiskers are activated.

Listing 11-1: Testing the Rear Whiskers

```
'-------------------------------------------
'testrw.bs2 - By Roger Arrick - 5/21/03
'
'This program tests the rear whiskers.
'When the whiskers are activated, the red LED
'turns on.

testrw: if in2 = 0 then testr    'Check rear whiskers
        high 10                   'LED off
        goto testrw               'Loop forever

testr:  low 10                    'LED on
        goto testrw               'Loop forever
```

Enter the program into the Basic Stamp editor software manually, or copy-and-paste the code from the Web site to the editor. Then download the program into ARobot and it will start executing immediately.

If you're unsuccessful at downloading, check the program for typographical errors, make sure the connection between your computer and the robot is working, and verify that the robot has good batteries and is turned on.

When Listing 11-1 runs, the red LED lights up to indicate that the whisker wire is touching the whisker bracket. If the LED is on continuously, make sure that the whisker wire is not touching the bracket. You may have to spend some time bending the whisker wire just right so that it hangs in the center of the bracket hole without touching the edges. If the LED never turns on, check the solder connection and verify that you used pin 3 from the body cable.

Make sure that the whisker bracket is insulated from the robot's base plate with nylon washers. Otherwise, the signal will be grounded.

Whiskers in reverse

If your cat was about to walk straight into a wall (or into a large German Shepard), you'd want it to back up, right? Well, your robot has the same capability now that it has whiskers. To help it retrace its steps if it encounters an object, you can use a simple program that works with the front and rear whiskers.

Listing 11-2 is a short program that shows how the rear whiskers can be used in navigation. The robot is commanded to move forward. If the front whiskers are activated, the robot moves backward. If the rear whiskers are activated, the robot moves forward.

Listing 11-2:   Navigating with the Front and Rear Whiskers

```
'------------------------------------------------------------
'navrw.bs2  - By Roger Arrick - 5/21/03
'
'This simple example program uses the front and
'rear whiskers to navigate.
'
'The robot will move forward until one of the following:
'  If one of the front whiskers activates,
'  the robot moves backwards.
'  If the rear whiskers activate,
'  the robot moves forward.

net     con 8               'Coprocessor network pin
baud    con 396             'Coprocessor baud rate

start:  serout net,baud,["!1R180"]      'Center steering
        pause 100

startb: if in14=1 then startb           'Wait 4 start button

main:   if in0=0 or in1=0 then mainf     'Front whisker?
        serout net,baud,["!1M116FFFF"]   'Drive motor forward
        pause 100
        goto main

mainf:  if in2=0 then main               'Rear whisker
        serout net,baud,["!1M106FFFF"]   'Drive motor reverse
        pause 100
        goto mainf
```

Make use of your new rear whisker sensor in future programs by using the code in Listings 11-1 and 11-2. Your robot will appear more intelligent and your cat might even learn a thing or two.

Now in and of itself this is not very useful behavior in a robot or a cat, but it will give you an example of how to read the whiskers and respond. The program comments show how the whiskers cause certain robot actions such as manipulating the steering or moving forward.

See the companion Web site (www.robotics.com/rbfd) for more elaborate navigation programs that use the whiskers.

Adding an Expansion Board

If you had a really huge robot — say the size of Rhode Island — there's no end to the accessories you could add to it. Sadly, most robots are not all that big, so there is some limit to how much stuff you can pile on.

Expansion boards provide a place for you to add more features to your robot. ARobot's controller board has a prototype area in the center where you can add small circuits and sensors, but it's not very big. In this section, you see how to add a large expansion board to your robot to make room for more projects.

Several of the projects in this book make use of this new expansion board.

ARobot's controller board has a special expansion connector that gives you access to all the important signals you need to run what you place on the expansion board, such as sensors, motors, or lights. A short cable connects the controller board's expansion connector to the expansion board. Power and ground are also provided.

The robot's base plate has a set of four holes (see Figure 11-8) in front of the controller board used to mount the expansion board. The hole pattern matches a common perf board and solderless breadboard sold by Radio Shack. (The perf board and breadboard are platforms for building circuits.) The hole pattern also matches that of the controller board, which allows you to stack the boards if you want to save space on the front of the robot for something else such as a gripper.

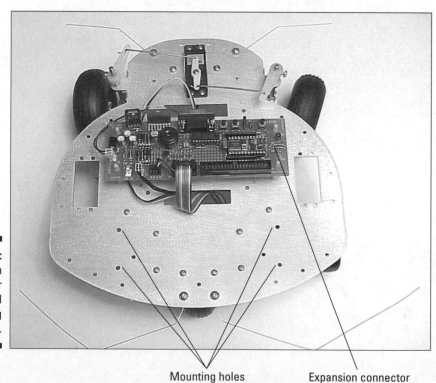

Figure 11-8:
Expansion
connector
and
mounting
holes.

Mounting holes   Expansion connector

Solving the cable conundrum

Buying flat cable from a distributor is difficult because it's usually available only in lengths of 10 or 100 feet and is often expensive. You have several other options for the 40-conductor flat cable.

✔ First, most computers use this type of cable for the hard drives, so you might have extra cables hanging around after computer upgrades. Check your junk drawer. If that doesn't work, ask your friends.

✔ Most computer stores have hard drive flat cables for sale. It doesn't matter if they're longer than you need because flat cable is easy to cut with sharp scissors.

✔ Flat cable with more than 40 conductors also works. You just cut out the extra wires.

Collecting the parts

Time for the robot scavenger hunt! Your first step in adding an expansion board is to gather all the appropriate parts. Table 11-2 shows the parts you need to install an expansion board and interface it to the controller. The spacers in the table are 1-inch long, but a shorter length will work as well.

Table 11-2	Expansion Board Parts List		
Part	*Vendor*	*Quantity*	*Approximate Price Each*
Perf board, 276-170	Radio Shack	1	$3.00
Spacer, 561-FSTP1000	Mouser	4	$0.50
Screw 6-32x1/4, 561-P632.25	Mouser	4	$0.07
Connector, 40-pin female, 517-8940	Mouser	2	$1.40
Connector, 40-pin male, 517-2540-6002	Mouser	1	$2.20
40-conductor flat cable	*	6 inches	Variable

*See the "Solving the cable conundrum" sidebar

Preparing the board

Take a close look at the perf board, and note that the bottom of the board has a series of connected holes. These are designed to help you wire circuits.

Some of these circuit traces need to be cut for the installation and use of the 40-pin connector. If you don't make this cut, many of the pins on the connector will be connected together, causing shorts.

The best way to make this cut is with an electric rotary tool, but you can also do it with a hobby knife. The cut should be made across 20 pairs of holes in one corner of the perf board. Figure 11-9 shows the traces being cut; look carefully at the position of the cut.

Installing the expansion board

Time to bring it all together by installing the expansion board. This process involves installing the connector, mounting the board, and making the cable that connects the controller board to the expansion board.

Installing the connector

Before you head off willy-nilly to install the connector, take time to do a little preparation. To prepare the board so that it accepts solder better, polish the copper traces with a pencil eraser or with steel wool. When you've finished polishing, clean off any debris, trying not to touch the copper with your fingers.

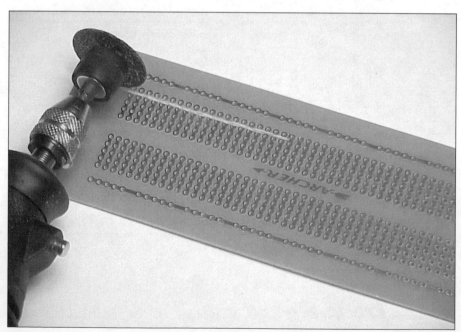

Figure 11-9:
Cutting
traces for
connector
installation.

Now install the connector following these steps:

1. **Position the connector, as shown in Figure 11-10.**

   Double-check the orientation and make sure the connector is pressed in all the way. If the connector has a keying slot, orient it the same way as the expansion connector on the controller board. (A keying slot is a specifically shaped hole that prevents you from plugging in the connector incorrectly.)

2. **Solder all 40 pins while watching for shorts.**

3. **Mark the top and bottom rows to identify them.**

   Notice that the expansion board has a row of connected holes along the top and across the bottom. You'll use these two rows to provide power to the expansion circuits. Following the example in Figure 11-11, use a marker to identify the top row as +5 and the bottom row as GND.

4. **Use a short wire to connect pins 1 and 2 of the expansion connector to the GND row and another short wire to connect pins 3 and 4 to the +5 row.**

If you need guidance on the pin numbers and their functions, refer to ARobot's user guide, which shows the connector pin-out.

The power distribution bus, which will carry +5 volts and ground to the circuits, is now complete.

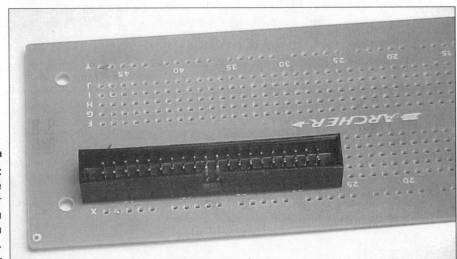

Figure 11-10: Position the connector before you begin soldering.

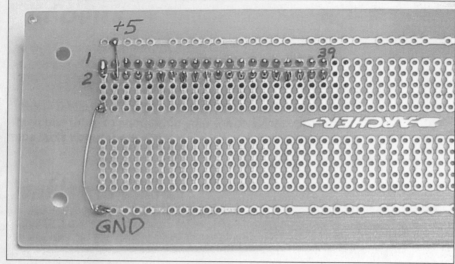

Mounting the board

An expansion board isn't much good if it's not connected to ARobot, so follow these steps to mount the expansion board:

1. **Locate the four mounting holes on the robot's base plate.**

   The mounting holes are located just in front of the controller board. They match the mounting holes for the expansion board.

2. **Orient the board so that the connector lines up with the expansion connector on the controller board, as shown in Figure 11-12.**

3. **Mount the four spacers using four screws.**

   The expansion board should fit perfectly onto the spacers. It may be necessary to drill out the four holes slightly on the perf board to accept the spacers.

Making the cable

Here's something you've probably been dying to try: The delicate art of flat cable construction. The goal is to make a short 40-pin flat cable to connect the controller board to the expansion board, as shown in Figure 11-13. Each end of the finished cable sports a 40-pin female header connector.

To shorten the cable, follow these steps:

1. **Cut the cable to length using sharp scissors.**

   If the expansion board will be mounted in front of the controller board, the cable should be 3 inches long; if it will be stacked on top of the controller board, the cable should be 6 inches long.

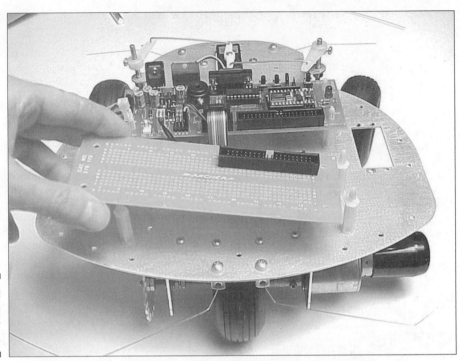

Figure 11-12:
Mounting
the expan-
sion board.

After the cut is made, inspect the cut edge and remove any strands of
wires that might be hanging out. The cut should be straight and
perpendicular to the edge of the cable.

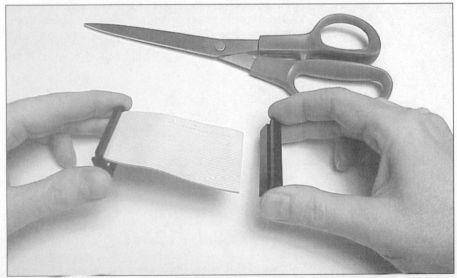

Figure 11-13:
Making the
cable.

2. **After deciding on the correct orientation, slide the connector onto the cable.**

   Header connectors often have keying tabs on one side and a pin 1 indicating arrow or triangle. Notice the keying notch on the controller board's connector. When complete, the connectors must match each other.

3. **If the cable has a colored edge wire, make that line up with pin 1 on the connector.**

   You should be able to see how the flat cable rests on top of the metal pins inside the connector.

4. **If you're making your cable from an existing cable that already has a connector, crimp one connector.**

   Connector crimping is usually accomplished with a special pressing machine but they're expensive and most people don't have one. Crimping can be accomplished in a standard bench vise or with a hammer. It's important that the connector stay on straight during the crimping process. If you use a hammer, place the connector on a flat surface and hold a small piece of wood on top of the wire and hit the wood gently.

   If the top of the connector breaks, it may still be usable by simply gluing it. You might want to practice on a few connectors first.

5. **After crimping, check that the connectors are straight.**

   If the connectors aren't straight, you might have a short. Test for shorts between adjacent pins with an ohmmeter and replace the connector if necessary.

Bringing it all together

You're almost finished. You still have to use the cable you just created to connect things. With the expansion board in place, attach the newly created flat cable from the controller board to the connector on the expansion board. Figure 11-14 shows the completed system.

Testing the board

Because there's no circuitry on the expansion board yet, the only thing you can test is power. Apply power to ARobot and use a voltmeter to measure the +5 volt signal at the expansion board. If all is well, download a program such as the wander program (refer to the "Running the Built-In Programs" section in Chapter 10) and test the robot for proper operation. At this point, your robot is ready and willing to accept new sensors, actuators, and more!

Building a Motorized Head

You know the old joke about how many people it takes to move a robot? Well, sad but true, if you add sensors or a camera to your robot, you have to position them manually by moving the robot. Wouldn't it be smarter to make a motorized head so that the robot can swivel it around into different positions all on its own?

Using a small head motor requires a lot less power than moving the entire robot and is much faster. (This is such a good idea in fact, that many organic creatures have moveable head assemblies too!)

Selecting a motor

Survey the motor landscape and you'll see many options for head motors, including stepper motors, DC servo motors, and model RC servo motors. Stepper motors would do the job but require a complex driver circuit. Traditional DC servo motor systems (like ARobot's drive motor) require optical encoder feedback and an expensive controller.

That leaves the simple RC servo motor, which you've probably seen used in model airplanes, cars, and boats. ARobot uses one of these standard RC servo motors to steer. These little jewels are perfect for small applications in which the range of motion is limited and control has to be simplified. They produce about 120 degrees or more of movement and are plenty powerful for our purposes. Most hobby stores offer RC servo motors for less than $20.

Figure 11-14:
The completed expansion board.

Mounting the motor

You just installed an expansion board, so why not use it? The head motor can be mounted anywhere on the robot, but I suggest mounting it to the expansion board because that allows the expansion board and head to come off as one assembly.

The disadvantage to using the expansion board for this motor is that it consumes space on the expansion board that could be used for other circuitry.

There are two options for mounting the motor to the expansion board:

✔ The simple option is to use double-sided tape. This tape is padded, strong, and easy to find at hobby stores, hardware stores, or even grocery stores. Just cut a piece of tape that fits the bottom of the motor and attach it to the front edge of the expansion board. Make sure the output shaft of the motor is centered on the robot's body so that the head is centered too. If you choose this option, you won't be needing the spacer and screws listed in Table 11-3.

✔ The more difficult option is to mount the motor using long metal spacers attached to the expansion board using screws. This results in a stronger attachment but requires some drilling. Table 11-3 shows the parts needed for this style of mounting, along with the perf board used for the head circuitry.

Table 11-3	Motor Mounting and Head Parts List		
Part	Vendor	Quantity	Approximate Price Each
Spacer 1.5", 534-2206	Mouser	2	$1.20
Screw 4-40x3/8, 561-P440.375	Mouser	4	$0.07
Perf board, 276-158	Radio Shack	1	$2.50

RC servo motors have four mounting holes, but I'm going to use only two of them to save drilling, some parts cost, and a little weight. Here are the steps to mount the RC servo motor:

1. **Mount a spacer on opposite sides of the motor on opposite corners.**

2. **Place the motor with spacers on the front edge of the expansion board so that the output shaft is centered within the expansion board.**

3. **Use a felt-tip marker to mark where the two holes should be drilled.**

4. **Drill the two holes.**

Be careful not to place the holes so that they interfere with the power traces on the expansion board.

5. **Mount the motor using the machine screws from the parts list.**

Adding a perf board

Now that the motor is mounted, you're ready to add the perf board, which will act as the head and give a space to mount sensors, cameras, and more. The perf board will be glued to a servo horn (*servo horns* are small plastic arms that attach to the motor's output shaft used for mounting), and then mounted on the motor.

You can purchase a bag that includes several styles of servo horns. Select a horn that's large and has a straight piece that crosses the shaft mount.

Follow these steps to add the perf board:

1. **If necessary, clip unnecessary pieces off the servo horn.**

   You need only a flat area to mount the head perf board.

2. **Trim the perf board to 3 inches by 2 inches.**

   To do so, use wire cutters or score the board with a hobby knife and snap the board on the sharp edge of a table. You can also use pliers to break off pieces.

3. **Attach the perf board to the horn by applying a bead of hot glue to the front and back of the perf board.**

   It's best to clamp the items so that they won't move during gluing. Use clothespins or molding clay to hold them together. Make sure that the position of the perf board allows access to the horn's mounting screw.

There are three tricks to working with hot glue. One, set up the items to be glued in advance. Two, don't let the hot glue touch your skin. Three, use a ceramic plate as a resting spot for the hot glue gun between uses.

Figure 11-15 shows the completed motorized head assembly. Pay particular attention to the mounting position on the expansion board. Note that the head is centered on the robot's body.

Controlling the motor

Time to put the motor to work. ARobot's controller board can control four RC servo motors. One motor port is used for the steering, and the other three are available for other projects. The Basic Stamp 2 controls motors by sending commands to the coprocessor, which does all the work. Lucky for you, the underlying complexities of motion control are hidden from view.

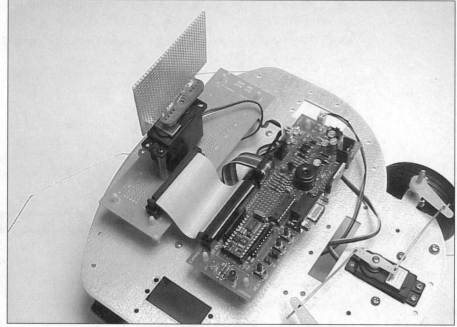

Figure 11-15:
An RC
servo motor
controls
the head
position.

To make the motor move, you must apply power and ground, and then supply a special control pulse that tells the motor which position to seek out.

Internally, RC servo motors contain their own gear train (sometimes called a *gear box*), driver, and controller. The motor's connector has only three pins:

✔ Pin 1 is control (usually white or yellow)

✔ Pin 2 is power (usually red)

✔ Pin 3 is ground (usually black)

There are exceptions to this pin-out. Check the packaging of the motor to verify its pin-out. The manufacturer's Web site will also have detailed information.

The four servo-motor connectors are on the controller board next to the body cable connector and are labeled J2, J3, J4, and J5. The first port (J2) is reserved for use by the steering motor. Plug the head servo motor into J3 and notice the colors of the wires. Normally the white (or yellow) wire is pin 1.

Listing 11-3 provides a simple program that you can use to test the head motor. After downloading it to the robot, the program simply moves the head from right to center to left. You can type this program yourself, or get it from the Web at www.robotics.com/rbfd.

If the head moves, you've succeeded. Now would be a good time to adjust the servo horn so that the head faces forward when the motor is in the center position. This will be important for future projects that require proper head positioning.

Listing 11-3:   Testing the Head

```
'----------------------- -------------------
'testhd.bs2 - By Roger Arrick - 5/21/2003
'
'This program tests the head motor by moving it
'from right to center to left.

net     con 8               'Coprocessor network pin
baud    con 396             'Coprocessor baud rate

testhd:
        serout net,baud,["!1R2FF"]  'Move head right
        pause 2000                  'Wait 2 seconds

        serout net,baud,["!1R280"]  'Move head center
        pause 2000                  'Wait 2 seconds

        serout net,baud,["!1R201"]  'Move head left
        pause 2000                  'Wait 2 seconds

        goto testhd                 'Loop forever
```

Troubleshooting

If — after following the instructions in the preceding section — the head just sits there and doesn't move, you have a problem. Check out the following possibilities:

✔ If the program doesn't download correctly, check the robot's batteries and make sure the power is turned on. Also check the serial cable connection.

✔ If the program download succeeds but the head motor doesn't move, check the motor connector orientation. Make sure the connector is completely pressed down on the pins.

✔ As a last resort, switch the connector around backwards and see whether that works.

Accessorizing

We accessorize everything these days: our cars, our outfits, even our cell phones. So why not add some neat accessories to your robot? Now that your robot has an expansion board, accessorizing becomes a snap.

In this section, I offer some ideas about how to make your robot stand out from the others by adding accessories that also provide some useful functionality. (No scarves, jewelry, and hanging fuzzy dice are involved.)

Wheel covers

A little dressing up can go a long way toward making your robot look cool. Taking a cue from sports cars, one of the first things you can do is add some flashy wheel covers.

Of course, it's possible to simply paint the plastic inner portion of the wheel, but you can also add a cover by using fender washers. Fender washers, which are available at hardware stores, have a large OD (outside diameter). The washers that fit ARobot's wheels should have a #10 hole and an OD of about 1 inch or 1¼ inch. Thin washers are best because they aren't very heavy.

Follow these steps to prepare and install the wheel covers:

1. **Polish or paint the washers before installing.**

   You may have to sand the washers first to get paint to adhere to the surface.

2. **If the inside diameter is too small to fit the axle, drill it to make it bigger.**

3. **Remove the collar that holds the wheel in place, slip the washer on, and then reinstall the collar.**

   Make sure the wheel can spin freely when you're finished.

You can see the snazzy look of these wheel covers in Figure 11-16.

Handles

If you're looking for items to spruce up your robot while providing some functionality, consider adding handles. It's difficult to carry and hold most robots, but handles can make the process easier while preventing you from damaging the robot by grabbing it by the wrong part.

Home improvement stores usually offer a large variety of handles. I suggest U-shaped units made of plastic or metal that are commonly used on kitchen cabinets. Plastic models are usually cheaper, weigh less, and come in a variety of colors. Metal handles are heavier but are stronger and look more high-tech.

Figure 11-16:
Flashy
wheel
covers
turn heads.

Handles require two mounting holes. The distance between the holes varies; 3 or 4 inches is common. When you buy the handle, #8 or #10 screws will be included. Follow these steps to install the handle on your robot:

1. **Find a good location on the base plate to mount the handles.**

   I suggest one handle on each side of the controller board, as shown in Figure 11-17, far enough away from the board to allow your hand to grab it easily.

2. **Mark the base plate where the holes should be and use a drill to make the hole.**

   Use a drill bit that's larger than the screw to provide extra room for the screw. Don't let metal flakes get on the controller board.

Beefing up the payload capacity

ARobot can easily carry several pounds of payload but there's always room for improvement. If you're considering adding some heavy accessories such as large batteries or a giant headlight to the robot, consider beefing up the chassis.

Figure 11-17:
Add handles for convenience.

The weakest link in carrying more weight is the rear wheel mounting brackets. They're fine for most applications, but with an easy change, they'll be much stronger. By simply adding a threaded rod between the two brackets, they'll be less likely to bend under stress. Follow these steps to add a threaded rod to the rear wheel brackets to strengthen them:

1. **Remove the wheels to allow access to the brackets for drilling.**

2. **Drill the holes so that the threaded rod won't interfere with the wheel axles.**

   Because the base is made of aluminum, it's easy to drill. Use a drill bit that's larger than the threaded rod, and don't let metal fragments get on the controller board.

3. **Install the threaded rod with nuts on the inside and outside of each bracket.**

4. **Tighten the nuts securely or use star washers to prevent loosening of the nuts as a result of vibrations.**

Figure 11-18 shows how the rod is mounted. I used a 10-32 threaded rod from a hardware store. It was 36 inches long, but I cut it with a hacksaw down to 5½ inches. You'll need four nuts, and you might have to sand the ends of the threaded rod to prepare it to accept the nuts.

Figure 11-18:
A threaded
rod adds
strength to
the wheel
brackets.

Running lights

Lights, camera, action! For some reason, lights just make for excitement, and a robot likes excitement as much as the next person . . . er, thing. So why not add lights to your robot?

I don't like to use too many lights because they draw current and reduce battery life, but there's no reason you can't add a few small lights to brighten your robot's day and convince onlookers that it's not just a bucket of bolts.

Be careful not to overdo it with lights because they can really hog the juice.

ARobot's power is approximately 12 volts, so many of the lights you find at automotive and motorcycle stores will work. They come in various sizes and mounting styles — some are rigid units and some are rope-like. The lights will receive power from the robot's controller board.

Here's how to mount lights on your robot:

1. **Mount the lights, using double-sided tape if possible, because that eliminates the need to drill holes and allows easy removal.**

   I suggest mounting the lights to the empty area on each side of the robot's base.

2. **Solder the + (positive) wire from the light, which is usually red, to the center pin of ARobot's power switch.**

    If there are just two wires of the same color, you can solder either.

3. **Solder the ground wire from the light, which is usually black, to pin 2 of ARobot's power connector (J9).**

Because most lights aren't polarized, the wires are interchangeable. This wiring will cause the running lights to come on when ARobot's power switch is turned on, as shown in Figure 11-19.

Clear dome

It's common for robots to have a clear dome covering their entire body. Personally, I don't like them because I often need to access something and a dome gets in the way. Nevertheless, a clear dome is an inexpensive method to protect your robot's innards while adding a high-tech look, as shown in Figure 11-20.

You can usually find clear plastic spheres in a variety of diameters at craft stores or hobby stores. They're often used as holiday decorations or as a container or lid. You'll need one that's approximately 10 inches in diameter and is split in half to form a dome.

Figure 11-19:
Brighten
your robot's
day with
running
lights.

Figure 11-20:
A dome can protect parts while looking high tech.

Remember, every part you add to your robot adds weight, which reduces battery life, so look for a dome made of thin plastic.

I won't kid you: Mounting the dome can be difficult. One common method is to use three or four small L brackets around the robot's perimeter. Drill small holes (not an easy task in itself) in the dome to allow for screws to be inserted into the dome through the brackets to connect them.

Another method is to have a single hole in the top of the dome that attaches to a threaded rod mounted in the center of the robot's base. You can then use a wing nut to secure the dome.

Making the dome work with the motorized head is another issue that you'll have to deal with. It may be necessary to carve out an opening in the dome for the head assembly to poke through.

Rubber bumpers

Robots commonly run into things — sometimes over and over. This could be funny, if it weren't for the fact that ARobot's base is made of aluminum and could possibly scratch a door frame, cat, or chair leg if it's going fast enough. Folks put bumpers on cars and chair rails in dining rooms, so it makes sense to add some cushioning to your robot's chassis.

A stroll through an auto parts store reveals many interesting accessory options for your robot, and one of those is garden variety plastic door trim. Made of flexible plastic or vinyl, its U-channel profile allows you to easily snap it on to the edge of a robot. The channel usually has a low temperature adhesive that keeps the trim from falling off. Various colors are available, including bright chrome and reflective gold. A package of 6 feet of trim costs 3 to 4 dollars.

Figure 11-21 shows trim added to all sides of ARobot, including the rear of its base.

Application is easy: Decide where to place the trim, cut it to length using wire cutters, and then press it on. Placement is not critical, but consider where your robot will be coming into contact with furniture and pets and add trim accordingly to cushion the blow.

If the trim has adhesive inside the channel, use a hair dryer to warm it up and help make it stick.

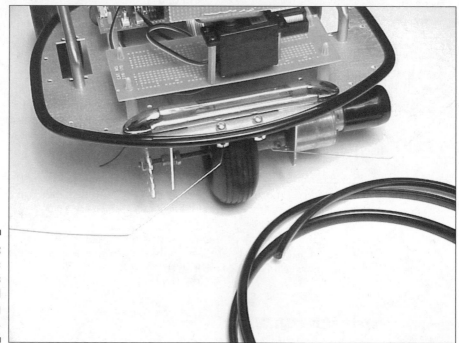

Figure 11-21:
Plastic trim makes a good bumping surface.

Chapter 12

Making Your Robot See the Light

*W*hen you walk from one room of a building to another, it's easy because you can use your senses — your vision, your sense of touch, and your sense of balance — to navigate quickly and efficiently. Now think about taking this stroll in complete darkness. You can't see the door until you bump into it (ouch!). The lack of light makes this ordinarily simple task difficult.

In the same way, any robot that you build must be capable of sensing its environment and gathering information to perform its tasks. Adding sensors to your robot is the best way to help it find out about its environment and give it the input it needs to make decisions.

One of the most useful sensors is the simple light sensor, also referred to as a *light detector*. The robot's controller can use the light-level information gathered by the light detector as input for navigation and even high-level actions, such as sending your robot into sleep mode when the lights go out to save power.

In this chapter, you find out how to build and set up a light-sensor system to show your robot the difference between night and day. Along the way, you pick up some nifty applications for this light-sensor technology in the real world.

Making Sense of Light Sensors

Adding a light sensor can be an eye-opening experience for your robot, but it's not like adding real vision. With a light sensor, your robot is like a human who's looking at the world through a sheet of paper. Sound useless? It's not.

Even when you're looking through paper, you can tell things about your environment, such as whether you're outside in the sun or in a dark room. Your robot can sense these things and more after you equip it with a light sensor.

A variety of commercial products use light sensors to trigger actions, including outdoor motion-sensing floodlights that turn off during daylight hours to save energy and automobile headlights that turn on automatically in low light. Also, many televisions use light sensors to automatically adjust the picture brightness based on the room light.

Your robot's light-sensing system has two main parts that work together: the hardware segment that does the actual sensing and the software segment that gathers and interprets the input that comes through the hardware.

The hardware part

Light-level sensors may do wonderful things for your robot, but good things don't always come in large packages. Most light-level sensors are small, two-wire electronic devices that change resistance as the light level changes.

In the world of electronics, *resistance* is a measurement of how much a given conductor blocks (or resists) an electrical current. Resistance is measured in ohms. You can look up more about basic electrical concepts and terms, such as resistance, in Chapter 5.

The amount of resistance at any given light level depends on the specific type of sensor you use. The CdS cell, which is named for the elements used in its construction (cadmium and sulfur), is the most common variety of light sensor. Figure 12-1 shows a common CdS cell that you can use as your robot's light sensor.

And the software part

The light sensor's change in resistance is what your robot senses. That is, the robot's controller can read the changes in resistance (resulting from light-level changes) and the program (the software part) can make decisions based on the light value readings. For example, your robot could regularly check the sensor, and if the resistance reading shows that the light is bright enough, it could then switch on solar panels to charge its batteries. That's accomplished with the software program.

The software used for this application consists of both a low-level program to interface with the hardware and a high-level program to work with the light-level value data.

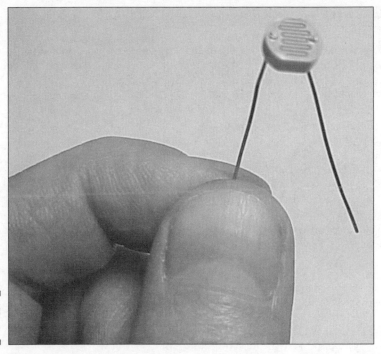

Figure 12-1:
A CdS cell.

Pulling Together the Light-Sensing Hardware

The hardware segment of your robot's light-sensor system has more than one piece. And these pieces must communicate and work together before the software segment comes in to make the robot do something. To complete the hardware segment, you need the following:

✔ **Something to sense the light:** This is a component such as the CdS cell shown in Figure 12-1. This electronic device acts as a variable resistor that gives off different readings under different levels of light.

✔ **Something to test the sensor:** You can connect a voltmeter to the sensor to check the sensor's resistance.

✔ **Something to collect readings from the light sensor:** This consists of the sensor in series with a capacitor.

✔ **Something to connect to the robot's controller:** Here's where you hook the circuit's components to the expansion connector or controller.

Read on to find out how to pull together these pieces.

Selecting the sensor

You can find light sensors at retail electronic shops and mail-order electronic supply houses, and you can even scavenge them from old equipment. Although you can spend a lot of money for something fancy, a $2 light-sensing cell is just fine for a simple robot project.

Following are some tips to keep in mind when you're looking for the right light sensor:

- ✔ **When using a catalog,** look for the section that contains optical components. Specifically, you want a CdS photo cell, often called a photo conductive cell. A bright-light resistance in the 1K to 100K range is best for this application.

- ✔ **When you're in a retail store,** look for CdS photo cells in the semiconductor section. Also, browse the product shelves, because the packaging often lists the specifications.

- ✔ **Some suppliers offer grab bags** with many types of cells. These are usually a bargain, but you're taking a small risk of not getting what you want. In your grab bag, you might find solar panels or phototransistors, which wouldn't work well with this project. Luckily, in this application, you can make almost any model of CdS cell work.

Testing the sensor before you commit

When dealing with anything electrical, a wise person tests the various components before putting everything together. Why go to the trouble of assembling everything just to find out that the first element you added was defective from the get go?

Follow these steps to test the sensor before you hook it up to your robot:

1. **Set your ohmmeter on the resistance (ohms) setting with a value of about 1M.**

2. **Connect the leads of the ohmmeter across the sensor.**

   Figure 12-2 shows you how an ohmmeter connects to a CdS light-sensor cell.

3. **Place the sensor near a bright light and write down the meter's resistance reading.**

   For the sensor I'm testing, the bright light resistance reading is 50K.

4. **Place your hand over the sensor (to simulate darkness) and write down the meter's new reading.**

   For the sensor I'm testing, the darkness resistance reading is .15K.

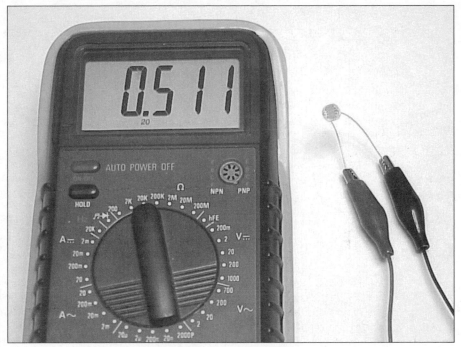

Figure 12-2:
Testing a
CdS cell
using an
ohmmeter.

These two readings — for bright light and darkness — will give you a general idea of the range of resistance you'll encounter as input for your robot's decision-making program. Normally, the resistance increases as the brightness increases.

Planning and building the circuit

Light sensors (CdS cells) act as variable resistors and can be used to generate a variable voltage. Therefore, microcontrollers that have an analog input port can easily read them because they're designed to read voltage levels.

If your controller, like mine, doesn't have an A/D (analog-to-digital) port — a port that can read analog voltages directly — you probably need a few extra components and some clever software to help your digital port read the input from the light sensor.

Figure 12-3 shows you the schematic for the circuit that I built to gather information from the light sensor and make it readable by my robot's controller. As shown, I constructed a simple voltage divider with a fixed resistor in series with the sensor across power and ground. As the light level varies, the voltage between the two resistors of this divider also varies.

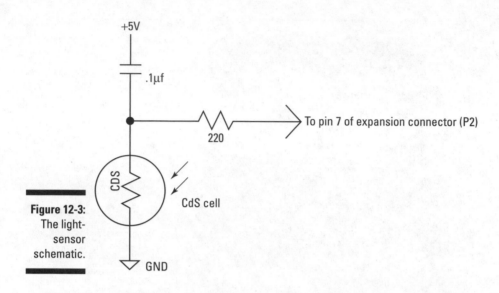

Figure 12-3:
The light-
sensor
schematic.

In this chapter's project, I use this circuit to read a change in resistance through a digital input port. (Read the sidebar "How the circuit works" if you want to know the nitty-gritty of how the circuit does its magic.)

After the low-level software reads the light-level values coming from the circuit, high-level software compares those values against preset values to help the robot make decisions. (In the section "Putting your Sensor to Work: Real-World Applications," I tell you about some surprisingly interesting and intelligent robot behavior that can then result.)

How the circuit works

You don't necessarily need to understand everything that's happening inside the circuit shown in Figure 12-3. But if you're interested, read on for the scoop. In the figure, the sensor is in series with the capacitor between ground and +5 volts. The signal is taken from the middle of this series of components and goes to the controller's digital port pin through a protection resistor. The digital port pin can read only an on or off condition, with a threshold of approximately 1.5 volts.

With this circuit, the program turns the port pin on, which discharges the capacitor, and then switches the pin to input mode, which lets the capacitor charge. When the charge reaches the threshold (1.5 volts), the software reads a logic-0 condition. The sensor's resistance determines how long it takes for the capacitor to charge. This amount of time becomes the light-level value. This scheme may sound complicated, but the circuit is simple and works great.

Time for a little hands-on experience with your circuit. Gather components that come as close as possible to the values shown in the schematic (Figure 12-3). Specifically, you need the following:

- CdS cell
- 220-ohm or 470-ohm protection resistor
- .1uf, capacitor

The components I used are shown in Figure 12-4.

Don't forget to notice the polarity of the capacitor. Capacitor polarity matters — a lot. A polarized capacitor that's used backwards can explode and scare your robot, so be careful! Also, a capacitor that's been installed backwards and doesn't explode tends to underperform and might fail. Not all capacitors are polarized, so look carefully for a plus sign (+) or minus sign (–) on the body. If you don't see either sign, the capacitor is probably not polarized and can be installed in any direction.

Build the light sensor circuit in the corner of your robot's expansion perf board by following these simple steps:

1. **Insert each component into the perf board by inserting their leads through the holes, as shown in Figure 12-5.**

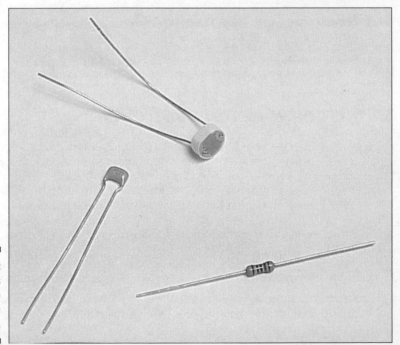

Figure 12-4:
The pieces you need to build the circuit.

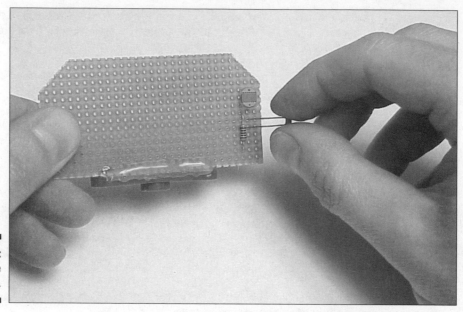

Figure 12-5:
Insert the
components.

Make sure that the components are close enough together so that the leads will reach each other, but leave the CdS cell leads long so that you can easily reposition the robot and point it in any direction (see Figure 12-6).

2. **Connect the component leads together on the bottom according to the schematic.**

3. **Solder the connections as shown in Figure 12-7.**

Interfacing to the controller

You've built the circuit but your robot is still in the dark. What's gives? Well, you need to make a few more connections before your robot can use the input from your light sensor. Specifically, you must connect the circuit to the power source, make sure the circuit is grounded, and hook the light signal to the controller. All of this happens on the bottom of the perf board.

Follow these steps to complete the light-sensing system's interface to your robot:

1. **Solder the ground signal to pin 1 of the expansion connector.**

2. **Solder the light signal to pin 9 of the expansion connector, which is P4 on the Basic Stamp 2 controller (see Figure 12-8).**

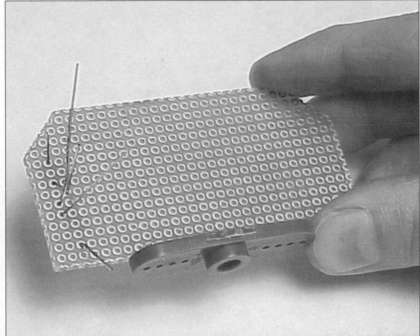

Figure 12-6:
The component leads protrude through the back.

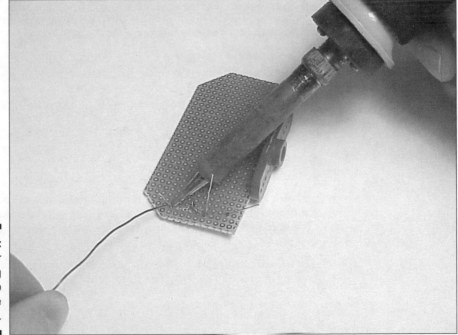

Figure 12-7:
Use your soldering skills to make the connections.

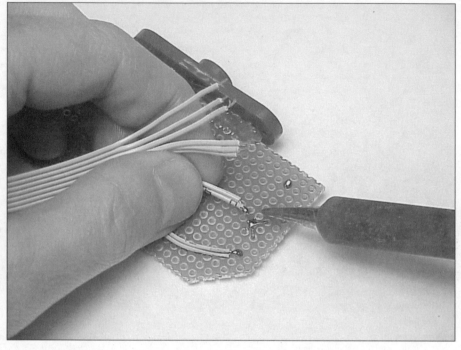

Figure 12-8:
Soldering
the wires to
the light
sensor
circuit.

3. **Check all connections for three dreaded problems that can plague electrical circuits: shorts, bad solder joints, and hardware interference.**

- **Shorts:** Beware of shorts. No, I'm not talking about a fashion problem; I'm talking about something much more troublesome — short circuits. Short circuits are a common problem when building any type of circuitry, especially when metallic hardware is involved. Before applying power to your new creation, verify that wires and components are not touching metallic hardware such as the spacers. Having wire or components touching metal where they shouldn't creates a short, which means at best that the circuit might not work and at worst that you might ruin a good part.

- **Bad solder joints:** Notice crusty looking joints. You want a shiny joint that covers the component leads with a good dab of solder. The component lead should protrude from the solder like a new tree in fertile ground.

- **Hardware interference:** Make sure that the components don't bump into anything nearby. The light sensor should have a clear view of the outside world.

4. **Use nippers to cut off pin 6 from SIP resistor RN1, and then reinsert the SIP resistor.**

A SIP resistor is a long slender black component that resides in a socket and is easy to remove. A SIP resistor is on each side of the BS2 on ARobot's controller board. RN1 is the one near the pushbuttons; RN2 is on the other side, near the expansion connector. The end of RN1 has a dot and is pin 1. Count from there to find pin 6. By cutting the pin, you remove a resistor that would otherwise interfere with the sensor circuit.

Engineers know that testing the circuit is the Moment of Truth. After you perform visual checks of the circuit construction and verify parts, you're ready to apply power to the circuit by installing batteries and turning on the power. When you do, absolutely nothing should happen because no software is yet in place to complete the light-sensing system. But you can check voltage levels in the circuit to make sure that all is well.

Writing the Software That Interprets the Hardware

Hardware is not enough to make your robot see the light. For that, you need a software program that tells your robot to gather data from the sensor you've attached to it.

The program I provide here is divided into two pieces: low-level software and high-level software. After you've downloaded these programs into you robot friend, it will be able to make sense of light and dark and return values for them.

This software must meet certain requirements. It must

- Gather data from the sensor
- Consume a minimal amount of memory
- Take a minimal amount of time to read the sensor
- Be modular and easy to use

Writing the low-level software

When I have a consumer complaint, I go to the top, but when it comes to programs, I prefer to start at the bottom. So, let's start with the low-level software, which is the part that deals directly with the hardware.

High-level software uses low-level software to communicate with the hardware, so the underlying complexity is hidden from the programmer. Modularizing code like this makes programming simpler, more portable, reusable, and easier to understand because after you get the low-level code working, you won't have to remember what's going on inside it.

For a light sensor, you need a routine to collect the light-level value, which will be in the form of a number in a variable. This subroutine will be available to your high-level program whenever it needs to know the light-level value. The code is simple because of the powerful rctime command that's built into the PBasic language inside the Basic Stamp 2 controller. (Aren't you glad the code is simpler than all this programmer jargon?) The subroutine is shown in Listing 12-1.

Listing 12-1:   Getting the Light Level

```
'-----------------------------------------------
'getll
'Reads the light level and sets variable lightl


getll:   high 4              'Discharge cap
         pause 1             'Short delay
         rctime 4,1,lightl   'Capture light level
         return              'Done
```

Hooking up with the high-level software

Moving up the ladder, it's time to write that high-level software to test the subroutine and the sensor itself. This program, shown in Listing 12-2, displays the value of the light sensor as a number on the screen once each second. For this to work, you must have your controller connected to the computer, as outlined in Chapter 8, and have the editor running, as described in Chapter 9.

Listing 12-2:   Testing the Light Sensor

```
'-----------------------------------------------
'testll.bs2 - Roger Arrick - 5/23/03
'
'This program tests the light-sensor hardware
'by constantly reading the sensor and displaying
'the results to the debug window.

lightl  var word            'Light Level

start:  gosub getll         'Get light level
        debug ? lightl      'Display result
        pause 1000          'Wait 1 second
        goto start          'Loop forever
```

It's a fact of life that programs are made to be tested, and now's the best time to do it. Type the testll.bs2 program into the program editor or get it from the Web site. This program should have the `getll` subroutine below it. Download the program into the controller and watch the debug window. Every second, a number is displayed in the window. Try exposing the sensor to various amounts of light and watch the value change. Cool, huh?

You'll find information about writing a program in Chapter 5 and details on downloading programs to your robot in Chapter 9.

This is the time to collect and record values for your future robot projects. If you don't live in a rainy spot such as Seattle or England, take your robot outside and get a sunlight reading, and then move inside and get readings with the lights on and with the lights off. You might also want a reading with a flashlight shining on the sensor for use in remote-control projects. Also make note of the direction that the values change. You'll probably find that the value gets lower as the light becomes brighter.

Troubleshooting the software

Having problems with writing or running the light-sensor program? That's entirely possible. If building and programming robots were totally trouble free, everybody would be doing it! Check this list of problems and potential solutions for a quick answer:

- **Download error:** This is usually an issue with communications between your PC and the robot's controller. Make sure you've selected the proper communications port. Also verify that the controller is getting power and that the computer cable is connected correctly.

- **No output:** Have you downloaded the program but no output appears in the debug window? Check the program carefully for typographical errors. Every character matters! Notice the difference between the letter *l* and the number 1, and the letter *O* and the number 0. Make sure that the editor window shows both the test program and the subroutine. Don't forget to download again after making changes to the program.

- **Value doesn't change:** If downloading worked and a value is displayed every second, but the value is always the same regardless of the light level, you could have a hardware problem with the sensor circuit or wiring. Turn the power off and then check all connections, component values, and component polarity. Also make sure that pin 6 is cut off from SIP resistor RN1, as described in the "Interfacing to the controller" section. A mistyped word in the program could also cause this problem.

- **Value doesn't change very much:** The sensor is probably working but it wouldn't hurt to make sure that there isn't a short or a component value problem. The total range of values is based on the type of sensor you have. As long as the value changes by 100 points or more between light and dark, the readings will be useful in measuring changes in light. Also make sure that pin 6 is cut off from SIP resistor RN1.

If you're still stuck, check the troubleshooting guide on our Web site at www. dummies.com/extras. In addition, you can find tested program files available for downloading at www.robotics.com/rfd.

Putting Your Sensor to Work: Real-World Applications

Great news — you can officially consider your light sensor a system now because you're finished with both the hardware and the software. That means you're ready to use the system in the real world.

It's time to dig right into some useful applications using your new light-level sensor system.

Making your robot sleepy

Did you know that most robots have insomnia? They just keep going with their mechanical eyes wide open until their power source runs out. With this simple program, you'll be putting an end to that sort of behavior.

The idea is simple: Your robot will regularly check the light level and go into sleep mode when the lights are off (that is, when it reads a specific light value). In sleep mode, the program will turn off the drive motor and conserve power. In wake mode, the robot will simply move forward. Listing 12-3 shows the program required to do all this.

Listing 12-3:   Putting Your Robot to Sleep

```
'-------------------------------------------
'sleepy.bs2 - Roger Arrick - 5/23/03
'
'This high-level program uses the light sensor
'to control wake and sleep modes for the robot.
'You'll have to adjust the trip point to work
'for your sensor.

lightl  var word              'Light level

        'If dark then just sleep here
main:   gosub getll            'Get light level
```

```
            if lightl < 100 then wake   'Trip point


            'sleep
            serout 8,396,["!1M1100001"]  'Stop the motor.
            sleep 1                'Sleep for 1 second
            goto main              'Loop forever

            'Wake
    wake:   serout 8,396,["!1M1139999"]  'Move forward
            goto main



    '---------------------------------------------
    'getll
    'Reads the light level and sets variable lightl

    getll:  high 4                'Discharge cap
            pause 1               'Short delay
            rctime 4,1,lightl     'Capture light level
            return                'Done
```

Running the sleep program

Make sure that you've entered the program code correctly, and then download it to the robot. The robot immediately begins reading the light level and decides whether to stay awake or take a snooze. In wake mode, the robot simply moves forward. In sleep mode, the robot stops and goes to dreamland (well, it really just stops).

You should set the trip point in the program according to the values you collected earlier. You might need to adjust the trip point to get the proper results. For example,

```
if lightl < 105 then wake
```

or

```
if lightl < 95 then wake
```

Remember that you have to set the all-important trip point to work with your sensor. Use the testll.bs2 program to determine a good trip point.

Although this program is not fancy, you can use the principles in more complex applications. One variation of this application is to set the trip point high enough that only a flashlight directed toward the sensor will allow the robot to wake. The result is a crude form of remote control.

Troubleshooting the sleep program

If you've come this far, you've already proven that your light-sensor system is working, so there's no need to look for hardware problems. Instead, the problem is most likely a typo in the program — perhaps a single incorrect character or an entire line. Printing a copy of the program might make finding the problem easier. Don't forget to download any changes you make to the controller.

Another thing to check is the trip value, which is used to switch between sleep mode and wake mode. You'll need to adjust that value depending on the response of your particular sensor. Use the testll.bs2 program and record the values for various trip-points that might be useful for other robot projects that need to sense various light levels.

If all else fails, check out the troubleshooting guide on the Web site at www.robotics.com/rfd.

Taking the next step

No doubt you want to find other ways to help your robot overcome insomnia. Try some of these ideas to expand the sleep program. Modify it to

- Turn on the red LED when awake (see Chapter 5 for details).
- Make the robot respond to whiskers when moving (Chapter 5 has program information for moving the robot and reading the whiskers).
- Set up a second sensor and use a flashlight to control forward and backward robot motion.

Programming a light alarm

In this project, you use the light sensor to sound an alarm when a light comes on. Use the application shown in Listing 12-4 to secure your room from intruders or to wake you when the sun comes up.

Listing 12-4:   Activating a Light Alarm

```
'-----------------------------------------------
'lalarm.bs2 - Roger Arrick - 5/23/03
'
'This high-level program uses the light sensor
'to activate an alarm when the light gets bright.
'You'll have to adjust the trip point to work
'for your sensor.

light1   var word              'Light level

         'If dark then do nothing
```

```
main:     gosub getll          'Get light level
          if lightl < 100 then alarm  'Trip point
          goto main            'Loop forever

alarm:    freqout 9.150.2000   'Alarm sound
          pause 100
          low 10               'Red LED on
          freqout 9,400,900    'Alarm sound
          pause 100
          high 10              'Red LED off
          goto main


'-------------------------------------------------
'getll
'Reads the light level and sets variable lightl

getll:    high 4               'Discharge cap
          pause 1              'Short delay
          rctime 4,1,lightl    'Capture light level
          return               'Done
```

Running the light alarm program

As you can see, the program is simple. It just reads the sensor repeatedly, checking for the trip point. When the trip point is reached, the alarm sounds. Detecting sunlight requires a different trip point than detecting room light, so the value in this program will be different than the one in Listing 12-3. After visually checking the program for errors, download it to the robot, and it will begin running immediately. Check the alarm trip point and adjust it if necessary. There is no wheel movement in this example.

Troubleshooting the light alarm program

If the program doesn't respond as expected, it's probably the result of a typo, so review the code closely. As with the preceding application, you'll have to set the trip value according to your sensor and the light you're reading. If all else fails, go to the book's Web site.

Taking the next step

Variety is the spice of life. Why not use your creativity to enhance the alarm program. Try these ideas for starters:

- Change the sound to emulate a telephone ringing by adjusting the parameters for the `freqout` commands.
- Make the red and green LEDs flash.
- Make the robot move forward and backward when activated.
- Program a timeout period so the alarm sound is limited to a single ten-second burst.

Positioning the light sensor

By positioning a single light sensor using a motor, you can scan an area for readings instead of moving the robot or using multiple sensors. This will allow your robot to find the direction of the light — and will add some exciting possibilities. Because your robot already has a motor-driven head assembly and your light sensor is already mounted there, all you need is some software to make the system come alive.

Controlling the motor

ARobot's controller board has the capacity to control four RC servo motors. One motor is used for steering and the other three are available to the robot builder. One of these is used for head positioning.

A special coprocessor on ARobot's controller board handles the four motor ports and eases the workload of the Basic Stamp 2. The complexity of controlling the motors is hidden from view. Your program controls the motor position by sending simple commands to the coprocessor using the `serout` command, which is built into PBasic.

Connecting the motor

Adding the head motor and head perf board is described in detail in Chapter 7. Before continuing, verify that the head motor's connector is plugged into the second position, next to the existing connector for the steering motor. The connector has three pins, and it's possible to plug it in incorrectly. Normally the first wire is white or yellow, the second wire is red, and the third is black. Use Figure 12-9 as a guide.

Scanning software

The first thing you'll need is a routine to scan and read the light sensor to collect several values. The routine in Listing 12-5 scans three positions: right, left, and center. This will tell you the general direction of the brightest light. After you collect the three values, the program can make decisions, such as to move towards the light or to avoid the light.

Listing 12-5:   Scanning and Reading Light Levels

```
'--------------------------------------------
'getls - Roger Arrick - 5/23/03
'Scans and reads the light levels at right, left, and center.

getls:
        serout net,baud,["!1R2FF"]   'Move head right
        pause 300
        high 4                       'Discharge cap
        pause 1                      'Short delay
        rctime 4,1,lightr            'Capture light level
```

```
serout net,baud,["!1R201"]        'Move head left
pause 600
high 4                            'Discharge cap
pause 1                           'Short delay
rctime 4,1,lightl                 'Capture light level

serout net,baud,["!1R280"]        'Move head center
pause 300
high 4                            'Discharge cap
pause 1                           'Short delay
rctime 4,1,lightc                 'Capture light level

serout net,baud,["!1R200"]        'Head motor off
pause 100

return                            'Done
```

With your scanning routine in hand, you can create a high-level program to collect light values, determine which is the brightest, and then move the robot in that direction. Listing 12-6 does just that. You can type the program or download it from the companion Web site. Place the getls routine below this program. Check out the robot in Figure 12-10.

Figure 12-9:
RC servo-
motor
connection.

Listing 12-6: Moving Towards the Light

```
'------------------------------------------------
'golight.bs2 - Roger Arrick - 5/23/03
'
'This program scans the light sensor using the head
'motor and moves the robot in the direction of the
'light.

lightl   var word              'Light left
lightc   var word              'Light center
lightr   var word              'Light right

net      con 8                 'Coprocessor network pin
baud     con 396               'Coprocessor baud rate

         low 9                          'Speaker off
         serout net,baud,["!1R180"]     'Center steering
         pause 100
here:    if in14=1 then here            'Wait for button


start:   if in0=1 and in1=1 then findi  'Stop if whisker
         serout net,baud,["!1M1100000"] 'Stop
         pause 100
         goto start                     'Wait for no whisker

findi:   serout net,baud,["!1M116FFFF"] 'Drive motor forward
         pause 100

         gosub getls                    'Get light level
         debug dec lightl,"-",dec lightc,"-",dec lightr,cr

         'Left brightest? (lowest value)
         if lightl < lightc and lightl < lightr then findl

         'Right brightest? (lowest value)
         if lightr < lightc and lightr < lightl then findr

         'Center must be brightest.
         serout net,baud,["!1R180"]     'Center steering
         pause 100
         goto start

findl:   serout net,baud,["!1R1D0"]     'Steer left
         pause 100
         goto start

findr:   serout net,baud,["!1R120"]     'Steer right
         pause 100
         goto start

'Place the getls routine here.
```

Figure 12-10:
Light
scanning
software in
action.

Running the light scanning program

The program simply scans and moves based on the light sensor readings. After downloading the program, disconnect the robot from the computer and set it on the ground or floor. Press button 1 to begin the program. The robot scans, looking for light, and then moves in the direction of the brightest light. This program works best in a slightly darkened room with one light on. Feel free to tinker with the steering values to make the robot respond more or less aggressively.

Troubleshooting the light scanning program

To troubleshoot the light scanning program — well, you know the routine:

- ✔ Look closely for simple typing mistakes. It takes only one little typo to make a system totally crash or have some bizarre behavior.

- ✔ Make sure that the head motor is moving the light sensors to the right, left, and center.

- ✔ Look carefully for the accuracy of the center position and adjust the head perf board if necessary.

- ✔ Don't forget to place the `getls` routine below the main program.

More Light Sensor Ideas

Just as the universe is expanding (and someday will go pffft), your ideas for how to use a sensor matrix can expand, too. In fact, there's no end to expansion options. To get you started, here are a couple of ideas on how to add functionality to the light-sensor system.

Improving direction sensing

Sometimes I think the robot builder's credo should be: If you have a good thing going, don't just sit there — make it better. Consider this idea for improving the capability of the sensors to detect direction: Use a small tube (see Figure 12-11) around the sensor so that it will detect light only from a certain direction. These tubes can be made of black paper, small plastic pipe, or heat shrink tubing, for example.

Don't use tubing made of metal because you could get a nasty short.

Attach the tubes using hot glue. The length of each tube will determine how selective the sensor is. (The longer the tube, the narrower the sensor's view.) Using a cone shape instead of a tube will let the sensor detect a wider area.

Software filtering

You'll find that the vagaries of the real world often get in the way of simple ideas. A good example is your new light sensor. All you really want is a good, stable reading, but sometimes your sensor gives you bogus values called *noise*.

Figure 12-11:
Sensor focusing tubes.

One method to deal with noise (both electrical and light in this case) is to use software filtering in your low-level code. For example, you know from your test program what range of values you should get from the sensor. That means the program can check for those limits, and take another reading if necessary. Following is the pseudocode to show you how this works. Pseudocode is a simpler version of the software that shows the flow and logic of the program without the hairy details needed to make it operate:

```
Read sensor
  If reading < lowlimit then reread light
  If reading > highlimit then reread light
```

Another trick you can use to ensure stable readings is to average them. The following pseudocode reads the sensor four times and calculates the average by using the formula you learned for averaging in third-grade math:

```
Smoothvalue = 0
For reading = 1 to 4
   Read sensor
   Smoothvalue = smoothvalue + reading
Next reading
Smoothvalue = smoothvalue/4
```

Overdoing it

You could spend your entire life just perfecting and polishing your light-sensor system — and trying the patience of everybody you know in the process. Try to resist the urge to get lost in a never-ending battle for perfection. It's sad to see a brilliant robot engineer get bogged down in the details and never finish a project. Be wise with your time and select a reasonable point to proclaim success; your robot will thank you.

Chapter 13

Some Like It Hot

Picture yourself in a dark room wearing earmuffs and a clothespin on your nose. Got it? Okay. On one side of the room is a red-hot woodstove, and on the other side is a 40-pound block of ice. Without your sense of hearing, smell, or sight, figuring out which side of the room the woodstove is on would have to be accomplished with your sense of touch — and more specifically, the part of your sense of touch that allows you to sense temperature.

A robot relies on senses, too, including that same sense of temperature. By providing your robot with a temperature sensor, you enable it to collect temperature data and make decisions based on that data. A robot might use temperature information to avoid dangerous places in a power plant or collect data for a remote weather station. Maybe in the future, a robot fire detector equipped with a temperature sensor will not only make a wailing sound when things get too hot, but also call the fire department and file a damage claim with your insurance company.

In this chapter, I show you how to build a simple temperature sensor that gives your robot one more piece of information about its surroundings. After you choose your sensor and put the pieces together, you'll write some software and do some "cool" experiments.

Sensational Temperature Sensors

Being human does have certain advantages, such as being born with something called skin, ready to sense temperature. Your robot wasn't born with such advantages, so if you want it to sense temperature, you have to provide it with a temperature sensor to recognize hot, cold, and everything in between.

Like other sensors, a *temperature sensor* produces an electrical change that can be measured by the robot's controller. Either a change in resistance (see Chapter 5 for more on resistance) or a change in voltage will do just fine as a signal to your robot friend that the temperature has shifted . . . that is, as long as your robot's programming understands what these changes represent.

To be meaningful, this electrical change must be proportional to the temperature being measured. For example, you might map a two-point change in voltage to represent a 1° change in temperature.

Temperature sensors come in two varieties: thermistors and semiconductor-temperature sensors. Each has its own features and benefits, which I explain throughout this section.

Thermistors can take the heat

A *thermistor* sounds like something that you'd use to spray your houseplants, but it's not. A thermistor (short for thermally sensitive resistors) is a kind of resistor designed specifically to measure temperature. You can see what one looks like in Figure 13-1.

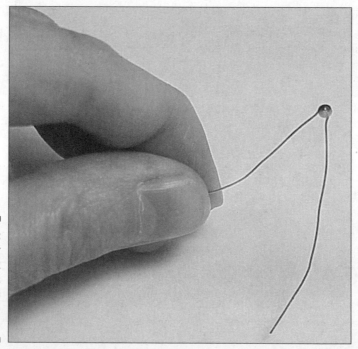

Figure 13-1:
A thermistor doesn't look like much, but it senses temperature changes.

Thermistors among us

Unbeknownst to you, thermistors are hiding in many modern-day electronic devices. They're often used to protect other electronic components, such as transistors, from overheating. When a thermistor detects overheating, the device can shut down a circuit to prevent damage or a fire. For example, if the power transistors in a stereo system's power amp overheat, the system is shut down until things cool off.

What's so special about a thermistor? Although most resistors change values slightly when the temperature changes, they don't change enough to be useful, and the response isn't calibrated or guaranteed by the manufacturer. A thermistor, on the other hand, is a resistor that is made to change resistance over a wide temperature range in a predictable and stable way.

Thermistors come in two varieties that you'll see listed in electronics catalogs and on thermistor packaging:

- **NTC:** Negative Temperature Coefficient. This simply means that the resistance decreases when the temperature increases.

- **PTC:** Positive Temperature Coefficient. You guessed it — the resistance increases when the temperature increases.

Thermistors are as common as houseflies; they can be purchased from most electronic retailers and mail-order dealers for about $2. And although NTC-type thermistors are the most common and often cost slightly less, the interface circuit on a robot can be modified easily to accommodate either type.

Sometimes you can purchase a grab bag of thermistors for a bargain price. Grab bags are great for experiments, but you'll need to do some testing of the response of the items before using them.

Semiconductors: Sensitive, three-legged creatures

Just as a new *Star Trek* series is spit out every few years, a new generation of sensors has become available that uses *semiconductors* instead of a resistive element such as a thermistor.

A semiconductor may contain many individual components that are sensitive to temperature and produce a voltage based on their temperature readings. Semiconductors are typically more accurate than their resistor counterparts.

Some semiconductor temperature sensors are even smart enough to transmit temperature data digitally to a local computer. These semiconductor-temperature sensors are an improvement because the output is usually a voltage that follows the temperature in a linear way.

I go on and on about semiconductors in Chapter 5; check it out if you need a quick refresher course.

Usually these sensors have three terminals, or legs: power, ground, and the output signal (see Figure 13-2). Because these semiconductor sensors have a voltage output instead of the thermistor's varying resistance, you can reduce the amount of circuitry needed to read the temperature. Engineers are always looking for ways to do more with less.

Figure 13-2:
A semi-
conductor-
temperature
sensor.

Semiconductor-temperature sensors are a bit more difficult to find than thermistors, but be persistent and you'll come across them. Dealers who specialize in electronic parts usually stock semiconductors (probably in a bin near the transistors and diodes).

Despite their benefits, these sensors don't cost much more than thermistors — only about $2.00.

Building the Temperature-Sensor System

I sometimes measure the difficulty of a project by how many bits and pieces are strewn around the table at the outset. I call this the *Amount of Stuff (AOS) standard*. By the AOS standard, the temperature-sensor system isn't a difficult project because the component count is small. And, after collecting these few parts, you simply do a little soldering, load the software, and you're finished.

Under the hood: Creating a temperature-sensor circuit

The circuit that you create when building a temperature-sensor system allows the controller to read the sensor's resistance by using the *capacitor charging method*. Most controllers, including ARobot's, can read only digital signals (one and zero) and are not well suited for reading analog signals, which vary between these values.

In the capacitor charging method, the sensor charges the capacitor, and the controller measures the time that it takes to charge. The result is an analog value. Although the resulting sensor value doesn't correspond to a specific temperature, you can take readings and make a map of values and temperatures. After you have a temperature reading, you can use that information in a program to create intelligent behavior in your robot.

For those of you who want to know the behind-the-scenes, techie stuff, here goes: The temperature sensor circuit sequence begins when the program you download to your robot to capture temperature data turns the port pin to a logic 1 to discharge the capacitor (see this schematic in Figure 13-3). Then the pin switches to input mode and waits for a logic 0. The capacitor begins to charge; the speed at which this charging occurs is determined by the resistance of the sensor. When the level reaches approximately 1.5 volts, the input pin trips to the logic-0 state. The program captures the time that it took for this to happen, and that result is proportional to the temperature. (Remember, the circuit will work even if *you* don't understand all these details.)

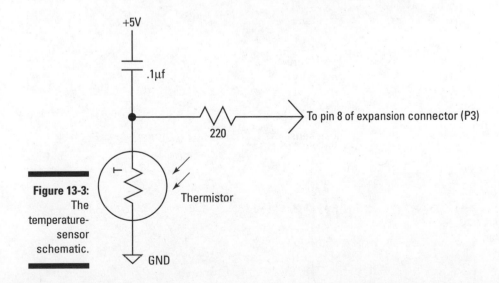

Figure 13-3: The temperature-sensor schematic.

+5V

.1µf

220

To pin 8 of expansion connector (P3)

Thermistor

GND

Collecting the parts

It's time to make room on your workbench and start gathering those parts to build your temperature-sensor system. Check out Figure 13-4 to see samples of each part so that you can identify them easily when you wander the aisles of your local electronics store.

Try to find parts that match the parts list shown in Table 13-1 as closely as possible. (You can visit the vendors listed in Table 13-1 at www.mouser.com and www.radioshack.com.)

Table 13-1	Temperature Hardware Parts List		
Part	*Vendor*	*Quantity*	*Approximate Price*
Thermistor 527-1006-50K	Mouser or Radio Shack	1	$2.50
Resistor 220 ohm, ¼ watt, 271-220	Mouser or Radio Shack	1	$0.02
Capacitor, .1uf (104), 80-C315C104M5U	Mouser or Radio Shack	1	$0.20

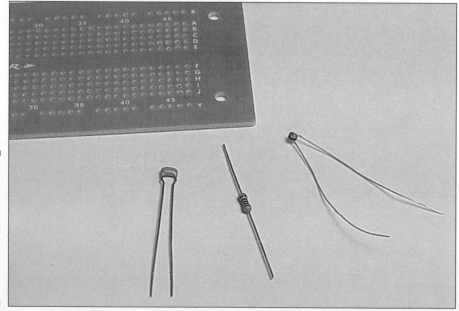

Figure 13-4:
Parts for the temperature-sensor circuit (left to right): capacitor, resistor, and thermistor.

Here are a few tips to keep in mind when you go shopping:

- ✔ The thermistor can be either NTC or PTC with a typical value of 25K to 250K.
- ✔ The resistor can be 470 ohm instead of 220 ohm.
- ✔ You can put resistors in series to add values.

Resistors can be purchased in various package sizes such as 5, 100, 1000, or even giant reels of 5000. The more you purchase, the cheaper they are. Retail packaging is the most expensive way to go and might cost you as much as 25 cents per resistor. At 1000 units, the price can drop as low as 1 cent each. I suggest that when you become hooked on building robots, purchase at least 100 of each item that you use frequently. At that quantity, resistors should cost somewhere around 2 cents each.

Installing the temperature sensor

Remember the first time that you touched a hot stove when you were just a tiny kid? You suddenly knew what *hot* was and never forgot the sensation. The next step in creating the same first experience of temperature for your robot is to install the temperature sensor. (No robots will be hurt in this process.)

Installing the temperature sensor is an easy, three-step process, but here are some guidelines to help you along:

- ✔ Select a location on the expansion board for installing the sensor and its associated components.
- ✔ Try to avoid placing the sensor next to other components that might get warm, such as the motor, making sure that the sensor's body isn't touching anything, including the perf board.
- ✔ The resistor and capacitor can be installed next to the sensor.

Follow these steps to install the temperature sensor:

1. **Insert the components into the perf board.**

   Placing the components as shown in Figure 13-5 allows the copper traces on the perf board to make the connections correctly. Use the schematic in Figure 13-3 as a guide to connect the component leads.

2. **Solder the component leads.**

Temperature sensor, meet the controller

Your robot has a temperature sensor, but without connecting it to the controller (the computer that runs the programs that control the robot), you could throw a pile of hot coals on it and it wouldn't know the difference. You have a little more work to do to make your robot temperature sensitive.

After the components are installed, you need to connect the power, ground, and signal from the temperature sensor to the robot's controller. Follow these steps to make these connections:

1. **Use a wire to connect the ground signal to the thermistor.**

2. **Use a wire to connect the +5v signal to the capacitor.**

   Refer to Figure 13-6 and the schematic (in Figure 13-3) as a guide.

3. **Wire the temperature signal to pin 8 of the expansion connector.**

   This connects the sensor circuit to P3 on the Basic Stamp 2 controller.

4. **Check all connections for shorts and bad solder joints.**

5. **Use nippers to cut off pin 6 from SIP resistor RN1, and then reinsert the SIP resistor.**

   A SIP resistor is on each side of the BS2 on ARobot's controller board. RN1 is the one near the pushbuttons; RN2 is on the other side, near the expansion connector. To find pin 6, count from pin 1, which is indicated by the dot at the end of RN1. By cutting the pin, you remove a resistor that would otherwise interfere with the sensor circuit.

Figure 13-5:
Building the circuit involves a few simple steps.

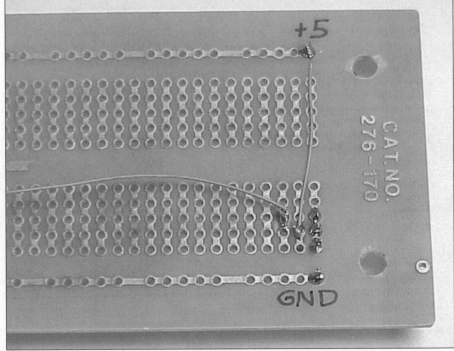

Figure 13-6:
The bottom
of the
perf board
after the
compo-
nents are
soldered.

6. **Mount the expansion board back on ARobot's chassis, and then connect the expansion cable.**

   See Figure 13-7.

Turning the switch

The final step is a little anticlimactic: When you turn on your robot, nothing special will happen at this point because it has no software component yet. (I get to that in the following section.) But it's still important to charge up the robot and do a simple test to see that power is getting through to the capacitor. Think of it as a sort of robotic dry run.

Don't forget to do the normal visual checks before applying power. Power for the temperature sensor will come through the robot, so simply turn it on when you're ready. Use your voltmeter on the 10- or 20-volt scale and check for 5 volts on one side of the capacitor. If you don't have 5 volts at the proper place, recheck the wiring and make sure your expansion cable is installed. That's all that you can test for at this point, so this mission is complete.

Figure 13-7:
The com-
pleted
circuit on
ARobot's
expansion
board.

The Brains of the Beast: The Software

A sensor without software is like a car without an engine. It might look good on the showroom floor, but it ain't gonna win any races. To get your robot on the temperature speedway, you need to add a program. This program consists of a low-level driver and a high-level program.

A major benefit of modularizing code with low-level driver software is that it hides the underlying complexity from the high-level software. It also allows you to make changes to the hardware without causing problems in other parts of the program. When the low-level code is finished and working, you won't have to think about it or consider what's going on inside it.

The driver at the lower level

Just like in the automotive world, to get behind the wheel in the temperature-sensor world, you need a driver. In this case, you have a simple low-level driver and a high-level program to test the driver.

In Chapter 6, I explain all about high-and low-level programs and why they work well in creating a modular application.

Your low-level driver software should meet certain requirements. For example, it must

- ✔ Gather data from the sensor
- ✔ Consume a minimal amount of memory
- ✔ Consume a minimal amount of time
- ✔ Be modular and easy to use

The low-level driver in this case is short and sweet: It simply reads the temperature sensor by using the powerful `rctime` command that's built into PBasic, returning the result as a number. The result can then be used by the high-level program to make decisions, such as moving towards the light or making an alarm sound.

The subroutine for the temperature-sensor driver is shown in Listing 13-1.

Listing 13-1:   Getting the Temperature

```
'-------------------------------------------------
'gett
'Reads the temperature and sets variable temp

gett:   high 3              'Discharge cap
        pause 1             'Short delay
        rctime 3,1,temp     'Capture temperature
        return              'Done
```

Taking things to a higher level

Now you need that simple high-level program to test the low-level routine and the sensor itself. This test program, shown in Listing 13-2, will display the value of the temperature sensor on the screen one time per second. For this program to work, the controller must be connected to the computer and the debug window must be open.

Listing 13-2:   Testing the Temperature Sensor

```
'-------------------------------------------------
'testt.bs2 - Roger Arrick - 5/23/03
'
'This program tests the temperature-sensor hardware
'by constantly reading the sensor and displaying
```

(continued)

Listing 13-2 *(continued)*

```
'the results in the debug window.

temp     var word            'Temperature

start:   gosub gett           'Get temperature
         debug ? temp         'Display result
         pause 1000           'Wait 1 second
         goto start           'Loop forever


'------------------------------------------------
'gett
'Reads the temperature and sets variable temp

gett:    high 3               'Discharge cap
         pause 1              'Short delay
         rctime 3,1,temp      'Capture temperature
         return               'Done
```

Hot or cold: Testing the software

Testing, testing, one, two . . . well, you get the idea. Here's where you find out whether these things that I've asked you to do throughout the previous sections of the chapter actually work. All that you need for this test is a little program and your own body heat.

Here are the steps to follow to perform the test:

1. **Type the testt.bs2 program in Listing 13-2 into the program editor or get it from the Web site.**

   This program should have the gett subroutine (refer to Listing 13-1) below it.

2. **Download the program into the controller and keep an eye on the debug window.**

   Every second, a new number is displayed in the debug window.

3. **Touch the sensor's body with your finger (or any other body part that you want to use).**

   The result should change because of the heat of your skin.

If your robot passed the test with flying colors, it's ready to collect temperature data. But keep in mind that the data it collects simply indicates *changes* in temperature. To translate those numerical changes into temperature read ings, you have to map the data to temperatures. I cover that in the next section.

If this test doesn't work (that is, the results in the debug window don't change), be sure to check out the "Troubleshooting temperature-sensor software" section for troubleshooting advice.

Making sense of the numbers

The results in the debug window won't do you much good if you don't know what they represent. You need to map the results to actual temperatures. This is also a good time to collect and record values for your projects. Here's how.

You can use a normal thermometer as a reference — you know, that one sitting at the back of your kitchen drawer under the stack of take-out menus. Place the thermometer near the temperature sensor but not touching it. Then escort the robot to various locations and record the result of the sensor and the thermometer's reading.

After you have several readings from places that vary in temperature (such as your toasty sunroom and your chilly basement), make a note of the way in which the reading changes when the temperature increases or decreases. The changes will vary depending on the type of sensor that you use.

You can test higher temperatures by using your hair dryer and lower temperatures by using the refrigerator. (This is handy in case you need to grab a snack after all your hard work on your robot buddy . . . make a quick sandwich and then melt its cheese with the hair dryer set on sirocco.)

Troubleshooting temperature-sensor software

When I was young and innocent, I believed that some things should work as advertised. After living through several versions of Windows software and buying gadgets advertised on TV, I'm now wise enough to know that that's not always true. In fact, even a simple project rarely works the first time.

If you encounter problems with the temperature-sensor process, check this list of possible problems for a quick fix:

> ✔ **Download error:** If you can't even get the darn program to download, you won't get very far. This is usually an issue with the communication between your PC and the robot's controller. Make sure that you select the proper communications port. Also, verify that the controller is getting power and that the cable is connected correctly.

✔ **No output:** You say that the program is downloaded but you don't have output in the debug window? Possibly you have what I call (in technical terms) *fumble fingers.* Check the program carefully for typographic errors. Really — one incorrect character could ruin your whole day. Note the difference between the letter *l* and the numeral *1* or between the letter *O* and the numeral *0* to see whether you might have typed the wrong one. Also make sure that the editor window shows both the test program and the subroutine. And don't forget to download the program again after making changes to it.

✔ **Value doesn't change:** The whole point of a temperature sensor is to . . . um . . . sense changes in temperature, so if the value that you get is a constant, you're missing the boat. If downloading worked and a value is being displayed every second, but the value is always the same no matter what the temperature is, you might have a hardware problem with the sensor circuit or wiring. Turn the power off, and then check all connections, component values, and the component polarity. Also make sure that pin 5 on SIP resistor RN1 next to the Basic Stamp 2 has been cut off (as described in the "Temperature sensor, meet the controller" section). A mistyped word in the program could also cause this problem.

✔ **Value doesn't change very much:** The sensor is probably working but it wouldn't hurt to make sure that there isn't a short or a component value problem. The total range of values is based on the type of sensor you have. As long as the value changes by 100 points or more between cold and warm, the sensor will work for these projects. Also make sure that pin 5 on SIP resistor RN1 has been cut off.

If all else fails, you're probably feeling bruised from running into brick walls. I don't want to think of you being bruised and battered. Instead of trying the same stuff for the tenth time, check the troubleshooting guide on the Web site, which will have some ideas that you might not have thought of. In addition, tested program files are available for download at www.robotics. com/rbfd.

Turning Up the Heat: Real-World Applications

Your spouse might tell you that tinkering with robots is a waste of time, but here's where you prove him or her wrong. Put your hard work to some good, practical use in the everyday world.

In this section, you play around with logging sensor data, setting up an alarm system to alert you if your robot is overheating, and controlling robot functions based on what your sensor senses.

The test program must be working before you can play around with these applications.

Your new hobby: Temperature logging

Sensor logging is used by scientists to collect data over a long period of time. This isn't because scientists don't have anything better to do — they just seem to like doing this stuff. Also this data helps scientists figure stuff out and make decisions about things such as the size of the universe.

Here's a list of some everyday sensor-logging applications:

✔ Monitoring seismic information for earthquake prediction

✔ Counting cars at an intersection to help program traffic lights

✔ Collecting weather data to help create forecasts

✔ Measuring water flow to minimize irrigation waste

Now you get to create your own sensor log. First, you get the robot to take sensor readings at regular time intervals and store the data in memory. Then, when the data collection is finished, the data can be uploaded to your PC and extracted and analyzed by a program that can create handy-dandy charts. (Charts and graphs are another thing that get scientists excited, so if you're the scientific type, this will make you very happy.)

The temperature-logger application in Listing 13-3 is something of a specialist program that collects just one nugget of information: temperature.

Listing 13-3: Logging the Temperature

```
'-----------------------------------------------
'tlog.bs2 - Roger Arrick - 5/23/03
'
'This program logs temperature data.
'Press Switch 1 to begin logging - 1 beep.
'Press Reset to stop - 2 beeps.
'Press Switch 2 to send the data to the serial port - 3
        beeps.
'Jumper J6 to erase the data - 4 beeps.

temp      var    word         'Temperature
beeps     var    byte         'Beep count
i         var    byte         'Misc
addr      var    word         'Address of EEPROM
samples   con    100          'Number of samples
```

(continued)

Listing 13-3 *(continued)*

```
storage data (samples)    'Storage for the samples

start:   beeps=2                  '2 beeps
         gosub beep
start1:  if in14=0 then log       'Start logging
         if in15=0 then send      'Send data
         if in13=0 then erase     'Erase data
         goto start1

log:     beeps=1                  '1 beep
         gosub beep
         for addr = storage to storage + samples
            gosub gett            'Get the temperature
            temp=temp*42/100      'Adjust value to 1 byte
            write addr,temp       'Save the value in EEPROM
            gosub delay           'Delay
         next
         goto start               'Done

delay:   for i = 1 to 10          '10 minutes
            pause 60000           'Delay 60 seconds
         next
         return

send:    beeps=3                  '3 beeps
         gosub beep
         for addr = storage to storage + samples
            read addr, temp       'Get the data
            debug dec temp, CR    'Send it out
         next
send1:   if in15=0 then send1     'Make sure switch off
         goto start1              'Done

erase:   beeps=4                  '4 beeps
         gosub beep
         for addr = storage to storage + samples
            write addr,0
         next
erase1:  if in13=0 then erase1    'Make sure switch off
         goto start1              'Done

beep:    for i = 1 to beeps       'Beep counting loop
            freqout 9,100,1500    'Beep
            pause 50              'Delay between beeps
         next
         low 9
         return                   'Done

'------------------------------------------------
'gett
```

```
'Reads the temperature and sets variable temp

gett:    high 3              'Discharge cap
         pause 1             'Short delay
         rctime 3,1,temp     'Capture temperature
         return              'Done
```

Here's an overview of how you use this program to collect and chart data:

1. **Download Listing 13-3 into your robot.**

   Now you can move the robot to a location, such as your doghouse, and start the monitoring process.

2. **The program collects the data (without disturbing your golden retriever) and saves it in the robot's memory.**

3. **Connect the robot to your PC and upload the data.**

4. **Open a spreadsheet program such as Excel and then cut and paste the data into a spreadsheet.**

5. **Use the spreadsheet program feature for creating charts and graphs to create various representations of the data to suit your needs.**

I hope it's obvious that you shouldn't try to measure extreme temperatures such as the center of a volcano or an iceberg at the Arctic Circle. After all, what good is a crispy or frozen robot engineer, anyway? Seriously, use your noggin and stay away from harmful places and temperatures — your robot creation will appreciate it.

Letting the temperature logger do its stuff

The actual process of logging temperatures is about as easy as falling off a log (and a lot less painful). Here are the steps that the logger goes through (and what you have to do to help it along):

1. **Download the Listing in 13-3 to begin the logger program.**

   You'll hear two beeps that indicate that the program has begun. At this point, the robot can operate autonomously — the cable to the computer doesn't even have to be connected.

2. **Erase the data memory by briefly placing a jumper on J6.**

   When you hear four beeps, the contents of the previous logging project or whatever random garbage existed from a prior program has been cleared. All data locations are set to zero.

3. **Press SW1 to begin the data collection process.**

   A single beep sounds to let you know the logging process has started. Data from the temperature sensor is read every 10 minutes (600 seconds) and then stored in memory. When two beeps sound, the data logging process is complete.

At any time before the process ends, you can press the Reset button to stop the logging process — you'll hear another two beeps. (This program is a little beep happy.)

The time between readings can be changed to once per minute or once per hour, but consider these important issues: Memory is limited to a few hundred samples, and the batteries must last long enough to do your logging. With a new set of batteries, you should be able to log temperature for eight hours or more. If you collect readings too often, however, you might run out of memory.

Going into analysis

If you collect data in the woods and nobody ever looks at it, is there a sound? Okay, enough philosophy. Data without analysis is like breakfast without a good slug of caffeine. After your robot collects data, it's time to analyze it.

Follow these steps to move the data to your computer and view it.

1. **Connect the robot to the computer with a serial cable.**

2. **Open the Basic Stamp 2 editor along with the debug window.**

   I use version 1.33.

3. **Press SW2 on the robot to make it send the collected temperature data to the computer.**

   The values appear in the debug window.

4. **Select all the data in the debug window by clicking and dragging over it with your mouse, and then press Ctrl+C to copy the data to the Windows clipboard.**

5. **Open a spreadsheet program, such as Microsoft Excel.**

6. **Select the first cell in the spreadsheet and then press Ctrl+V to paste the data in the spreadsheet (see Figure 13-8).**

7. **Save the spreadsheet file.**

With the data in your spreadsheet, you can manipulate it in many ways. Most spreadsheet programs offer a graph function. For example, in Excel, you select the data you want to include in a chart, and then choose Insert⇨Chart and follow the simple instructions to select a type of chart and various data and formatting options.

You might want to use a line chart format, like that shown in Figure 13-9, for example, to see the temperature trend over the entire logging time.

Troubleshooting the logger

Like any program, every character in the logger program matters. If you've made a mistake in the code, the editor will reveal most errors during download, but it doesn't always catch them all. If all else fails, download the tested source code from the Web site at www.robotics.com/rbfd.

Danger: Temperature alarm

"Danger, danger, Will Robinson." Okay, I've wanted to say that since I started this book, and now I have it out of my system. If you want your robot to run around sending out as many alarms as a series of *Lost in Space* episodes, have I got a program for you. You're about to discover how to program a heat alarm.

Programming a heat alarm is easy. This program constantly checks the temperature sensor and compares the value with a *trip point* set by the programmer. When the trip point is triggered, an alarm sounds. If you want to get creative, you can program the robot to perform any action upon tripping, such as moving to another location or flashing lights.

In the short program shown in Listing 13-4, notice special code for different trip-on and trip-off points. This separation of trip-on and trip-off points is called *hysteresis,* which prevents a frequent on/off action of the alarm. This is the same concept used in your house thermostat to prevent constant cycling.

Figure 13-8:
Put your temperature data in a spreadsheet to create a variety of charts or graphs.

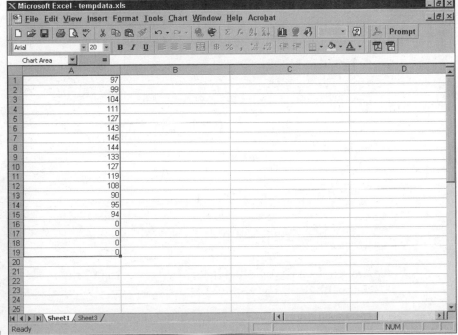

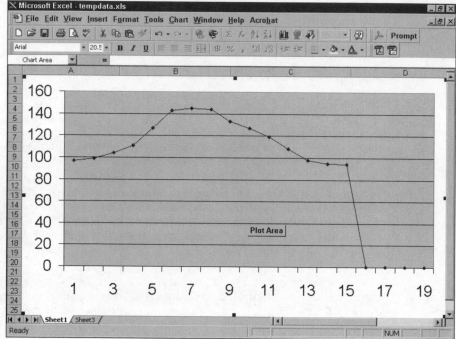

Figure 13-9:
Create a temperature trend line graph with Excel's chart feature.

Listing 13-4: Sounding a Temperature Alarm

```
'-----------------------------------------------
'talarm.bs2 - Roger Arrick - 5/23/03
'
'This program sounds an alarm when the temperature
'trip-point is reached.


temp      var   word                   'Temperature
triph     con   500                    'High trip point
tripl     con   100                    'Low trip point

start:    freqout 9,100,1500           'Beep
          pause 50                     'Delay between beeps
          freqout 9,100,1500           'Beep
          pause 2000                   'Delay before starting

again:    gosub gett                   'Read the temperature
          if temp > triph then alarm   'Check for on trip
          goto again                   'Continue forever

alarm:    low 10                       'Red LED on
```

```
        freqout 9,500,800        'High beep
        pause 100                'Delay
        high 10                  'Red LED off
        freqout 9,500,300        'Low beep
        pause 100                'Delay
        gosub gett               'Read the temperature
        if temp < tripl then again 'Check for off trip
        goto alarm


'-------------------------  ----------------
'gett
'Reads the temperature and sets variable temp

gett:   high 3                   'Discharge cap
        pause 1                  'Short delay
        rctime 3,1,temp          'Capture temperature
        return                   'Done
```

The main difficulty with this program is setting the trip point. The best advice that I can give you is to experiment until you get the result you want. You might also have to experiment with the spread between the on-trip and off-trip points to make the response useful. Also, make sure the greater-than (>) and less-than (<) signs are correct for your particular temperature sensor (see these in the if statements in Listing 13-4).

Robots that run hot and cold

This application is for those of you — and I know you're out there — who want to see your robot run around in circles, flash lights, and whistle *New York, New York.*

More ideas for the logger

Your temperature data logger will perform quite well when you attach it to the robot base. All you really need is the controller board and your temperature circuit. This means that your data logger can be carried easily to remote locations for temperature monitoring.

How about putting the logger in your attic to see how hot it gets up there during the day? Or try putting the logger in the garage to see how cold your Chevy gets at night. Placing your logger in a freezer or a car trunk are good experiments (and way better than locking yourself in the trunk with a thermometer).

The thermistor can also be connected to the circuit with long wires, which allows you to measure temperatures in hard-to-reach locations, such as in a glass of iced tea, under the dishwasher, or on top of Mt. Everest. Actually, forget Everest — there's a limit to the acceptable length of the wires. You should, however, be able to get away with 48" or more.

More alarming ideas

The heat alarm concept can be carried over to security applications, which will be a common job function for robots of the future. A central alarm and a sprinkler system could be activated by using a wireless communications link, for example. Integration with a cell phone would allow a robot to contact the fire department when the heat in your house reaches a certain temperature.

Although these might sound like futuristic and ambitious goals, all the technology exists to do these tasks and more. It's just a matter of integrating that technology into a reliable robotic system.

The short application in Listing 13-5 shows you how to use temperature data to control various functions of your robot, in this case, sound. You can place your finger on the sensor to change the sounds. The gett routine returns a number that corresponds to temperature, so you can use that number to control sounds, lights, and even the robot's motion.

Listing 13-5:   Using Temperature to Control Sound

```
'-------------------------------------------
'tsounds.bs2 - Roger Arrick - 5/23/03
'
'This program uses temperature to control sound.
'Experiment with the snd constants to change response.

temp    var    word                    'Temperature
ontime  var    word                    'Sound on time
sndf    var    word                    'Sound frequency
delayt  var    word                    'Delay between sound

snd1    con    2                       'Time on constant
snd2    con    10                      'Frequency constant
snd3    con    1                       'Delay constant

start:  freqout 9,100,1500             'Beep
        pause 50                       'Delay between beeps
        freqout 9,100,1500             'Beep
        pause 2000                     'Delay before starting

snd:    gosub gett                     'Read the temperature
        temp=temp*42/100               'Adjust value to 1 byte
        ontime = temp*snd1             'Sound on time
        sndf = temp*snd2               'Sound frequency
        delayt = temp*snd3             'Delay between sound
        freqout 9,ontime,sndf          'Make the sound
        pause delayt
```

```
        goto snd                    'Continue forever


'------------------------------------------------------
'gett
'Reads the temperature and sets variable temp

gett:   high 3              'Discharge cap
        pause 1             'Short delay
        rctime 3,1,temp     'Capture temperature
        return              'Done
```

Not getting the sound that you want? Tinker with various settings for the snd1, snd2, and snd3 constants to change the sound response. The thermistor will respond slowly, so you'll just have to amuse yourself while you wait.

You can easily adapt this program to respond to other sensors, such as the light detector described in Chapter 12. Just change the value 3 in the gett routine to read the light sensor instead:

```
gett:   high 4              'Discharge cap
        pause 1             'Short delay
        rctime 4,1,temp     'Capture light
        return              'Done
```

Then use a flashlight to control the sound and have a blast in the dark!

Chapter 14

Halt! Who Goes There?

● ●

In This Chapter

▶ Understanding how motion detectors work

▶ Adding a motion detector to your robot

▶ Writing software to control a motion detector

● ●

*W*atch a baseball game sometime and try to figure out how a pitcher senses motion when the guy behind him is trying to steal third base. The pitcher probably uses his sense of hearing, as well as peripheral vision, to pick up clues that somebody is headed for third who should be staying put on second.

Now imagine how handy it would be if your robot could sense motion, just like that pitcher. It won't become the next Sandy Koufax, but it could react to motion and make decisions, such as alerting you to danger or backing up when something approaches it.

In the project in this chapter, I show you how to add a motion detector to your robot. The details of this project are specific to ARobot and the software is designed for the Basic Stamp 2 controller, but you should be able to apply this information to other platforms and controllers as needed.

After a short introduction to motion-sensing technology, you jump right into the process of adding the sensor. You finish the chapter with some fun applications. Play ball!

Detecting Motion: An Overview

Your first experience with a motion detector may have been the time you went to your friend's house for dinner and a blinding light clicked on as you approached the front door, causing you to fall in the ornamental bird bath and drop the pimento dip you were carrying. Here's what happened: The light had a motion detector attached, you moved within the range of the motion detector, it caused the light to flash on, and you took a bath.

The term *motion detector* includes any sensor that's capable of triggering some action in response to motion, such as turning on a light. The most common type of motion detector is the *passive infrared (PIR) sensor,* which is the type that startles you outside houses at night and the type you use in a robot (see Figure 14-1). Other varieties of motion detectors include light sensors, cameras, and even mechanical sensors such as trip wires and floor mat switches.

In the last 20 years, PIR sensors have become commonplace and inexpensive. You've probably seen PIR sensors mounted on indoor walls of houses or stores as part of an alarm system. Many houses now have outdoor lights that are triggered by moving objects for safety and security. PIR sensors can be used also to trigger a camera in a surveillance system.

Improvements in technology have reduced false triggering, making motion detectors much more reliable than they once were.

Unlike people, who sense motion with their sight, PIR detectors operate by sensing heat. As you can imagine, simply sensing the presence of a heat source wouldn't be very useful because the sensor would simply stay on if you turned up your thermostat or threw another log in your woodstove. The beauty of modern PIR sensors is their internal multiple-sensor configuration, a special lens, and an intelligent controller; together these make up a system that detects only objects that move across the sensor's field of "vision."

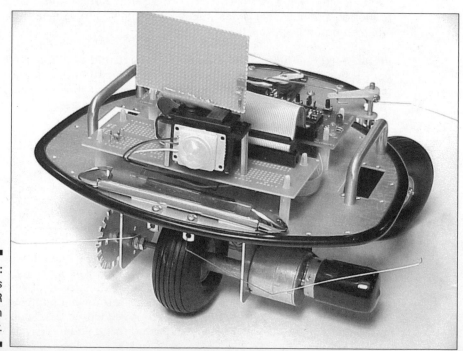

Figure 14-1:
ARobot's
new PIR
motion
detector.

Most PIR detectors have two or more separate areas that can detect heat. A lens directs the heat information onto the surface of the sensor and may also filter certain areas and certain types of light to limit false triggering. A small controller circuit reads the sensors and attempts to determine whether an object (usually human or canine in my house) has moved. If motion is detected, a signal is produced that can be used by the high-level system — in our case, our robot friend — to cause a response, such as setting off an alarm.

Luckily, intrepid robot builders don't have to construct motion-detector systems from scratch. The complexity of the PIR detector is self-contained in a small, ready-to-use module. You need to simply supply the sensor with power to operate and to produce an on-off trigger signal.

Building the Motion-Detector System

Your robot is probably just sitting there, waiting for that moment when it will be able to detect others moving around it (hey, it's lonely in there, you know), so we'd better get started.

Collecting the parts

The first step is to go on a scavenger hunt for the stuff you need for this project. You'll be glad to hear that the parts for your motion detector are inexpensive and fairly easy to find. You'll need the sensor module itself, some double-sided tape, and some wire (see Figure 14-2).

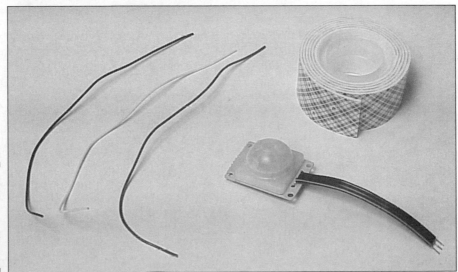

Figure 14-2: Parts used to build the motion-detector system.

You can find motion detectors built into products such as automatic light controls. If you have nothing better to do with your time than to reverse-engineer these products and you have the technical ability to do so, they're an option.

For most hobbyists, the best solution is to find a ready-made module sold on its own and designed to operate on low voltages. A casual stroll through your local home-improvement store will reveal one or more motion-detector modules lurking on the shelves. However, these are usually designed to operate with an alarm system and require *high* voltages.

I'm happy to report that COMedia Ltd. in Hong Kong makes the KC7783R PIR module, which is exactly what you're looking for. At 1 by 1⅜ inches, it's small enough to use almost anywhere, including in your robot. With a voltage requirement of 5 to 12 volts, it's a perfect match for your robot's power consumption needs. As Figure 14-3 shows, the module comes complete with the sensor, lens, and a controller, which means you have to perform only a minimal amount of work to implement this fine device in your robot.

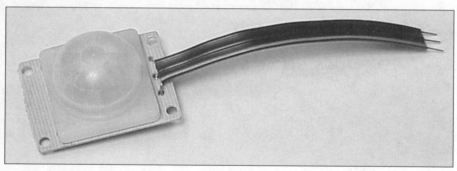

Figure 14-3:
The
KC7783R PIR
detector.

Purchasing this type of motion detector directly from the manufacturer is impractical due to shipping charges and minimum order requirements, but several companies sell them at a reasonable price.

Marlin P. Jones has been in the electronic component business for many years and offers this part as #7860-KT. The modules are less than $10 each and come with a small data sheet that describes their operation, specifications, and wiring. Ask for a catalog by contacting Jones online at www.mpja. com or by calling 1-800-652-6733.

Because the PIR sensor you use requires only power and creates a simple output signal, wiring is minimal. In fact, this project is so simple, the majority of the work involves mounting the sensor.

You can use almost any type of wire between the sensor module and the expansion board. Wire is either stranded or solid. Stranded wire is made up of many strands of smaller wire twisted like a rope. It's strong, flexible, and usually pre-tinned, which makes soldering easier. Solid wire is usually untinned copper, and because it's so stiff, it will break if bent repeatedly. I suggest you use stranded wire for this project.

You need only a small piece of double-sided tape, but you'll have to buy a complete roll. Or instead of using double-sided tape, you could use a Velcro-type fastening material with adhesive backing. This would allow you to easily remove the sensor if needed.

If you buy a roll of double-sided tape, save the remainder in a plastic bag to prevent it from drying out.

Table 14-1 lists the parts with possible sources and prices.

Table 14-1	Motion Detector System Parts List		
Part	**Vendor**	**Quantity**	**Approximate Price Each**
PIR Sensor Module, 7860-KIT	Marlin P. Jones	1	$10.00
Double-sided tape	Hardware store	1"x1"	$2.00
Wire, any type	Mouser or Radio Shack	8"	$1.00

Wired for motion

I find it best to do the wiring before mounting the sensor. There are only three wires: +5 volts, ground, and the sensor-output signal. Wiring consists o attaching three 6-inch sections of wire from the sensor to the expansion board. My sensor came with a 2-inch section of wire already attached, but that was too short for my liking, and using new wires is easier than extending existing ones.

It's always best to have a little extra wire length to help with future maintenance and repairs.

The sensor will come with a data sheet that shows the pin-out. Without the pin-out, there's no way to know which signal is which with certainty; and guessing will surely lead to a damaged part. If you didn't get a data sheet, contact the vendor, who may refer you to a Web-based document or mail the sheet to you. Figure 14-4 shows the sensor pin-outs of the unit I purchased. Yours may differ.

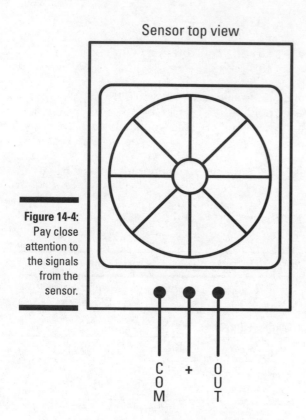

Sensor top view

C
O
M   +   O
U
T

Figure 14-4:
Pay close
attention to
the signals
from the
sensor.

Little current is involved with this sensor, so you can use small, flat cable wires that have been peeled off a larger cable. You can also use larger wires such as awg 22 or awg 24 that might be lying around from a previous project. In fact, almost any wire will do. If you want to color-code the wiring, use red for +5 volts, black for ground, and some other color such as white for the output signal.

Follow these steps to complete the wiring:

1. **If the sensor comes with wire already attached, remove it by using your soldering iron to heat up the joint (see Figure 14-5).**

2. **Cut the three wires into 6-inch lengths.**

3. **Strip and tin each end with a small amount of solder, but not longer than about $\frac{1}{16}$ inch to reduce the possibility of shorts.**

   It takes only a small amount of exposed wire to make a connection.

4. **Solder the wires onto the sensor, as shown in Figure 14-6.**

   Double-check for good solder joints and for shorts.

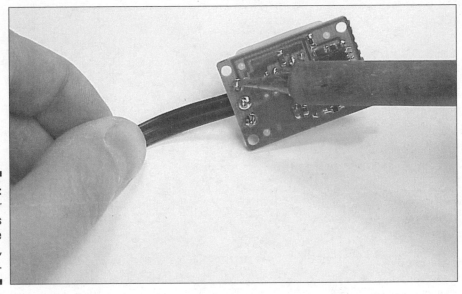

Figure 14-5:
If your sensor has a wire attached, remove it.

5. **Solder the +5 volt wire to the +5 volt rail on the bottom of the expansion board.**

The sensor will be mounted up front on the body of the head servo motor, so make sure the wire is long enough to reach.

When attaching the three wires to the expansion board, you have to be careful which wire is which. Frequently check your wiring against the data sheet.

6. **Solder the ground wire to the ground rail.**

7. **Solder the output signal to pin 10 on the expansion connector (see Figure 14-7).**

It's important to correctly count the pins on the expansion connector. Solder shorts usually happen on the expansion connector because the pins are so close together.

Figure 14-6:
Attach the new wires to the sensor.

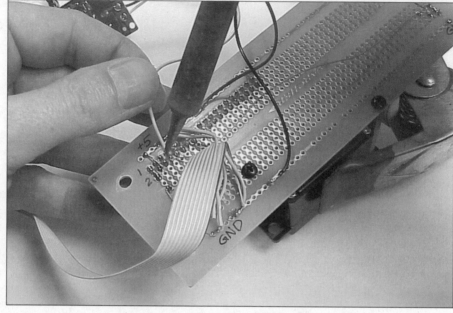

Figure 14-7:
Connect the
output
signal to the
expansion
connector.

Figure 14-8 shows the sensor in front of the robot but not mounted.

Testing the motion detector

A motion-detector sensor that doesn't sense motion is about as useful as a lottery ticket with a losing number on it. It's important to test the sensor to make sure everything is operational. This is a good time to test the sensor, because testing before the sensor is mounted on the robot makes it easier to access the wires behind the sensor, if you need to.

Follow these steps to apply power to the sensor:

1. **Make sure the expansion board is mounted to the robot and that the expansion cable is attached to the robot's controller board.**

2. **Apply power.**

   It's important that the sensor's electronics and exposed wires don't touch any metal or other circuitry. Otherwise, you could damage the components or your batteries.

3. **Use your voltmeter on the 10- or 20-volt DC scale and check that +5 volt power is making it to the sensor.**

   Don't let your probe slip — damage might occur to the sensor.

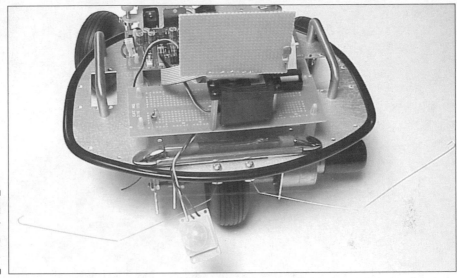

Figure 14-8:
The sensor
is ready to
be mounted.

You can also check the sensor's operation at this point by following these steps:

1. **Place the black probe on a ground point anywhere in the system.**

2. **Place the red probe on the motion detector's output signal (see Figure 14-9).**

3. **Move your hand in front of the sensor.**

   You should notice the signal moving between a low voltage (less than 1 volt) to a high voltage (greater than 4 volts) as you move your hand in front of the sensor.

Give me a hand

A robot might have eight legs, but sadly, humans only have two hands. That's a problem because we often need three or four to get the job done. This is especially true when working with voltmeter probes.

For example, when testing a motion detector, you need two hands to hold the probes and another to make a motion, such as waving in front of the sensor.

A simple solution is to use wires with alligator clips on each end. They're available from most electronic shops and usually come in a package with several color clips. Clip one side to your probe and another side to the circuit. Look ma, no hands!

Figure 14-9:
Using alligator clamps aids testing.

Most PIR sensors have a brief warm-up period of a few seconds before they become operational. You may have to remain very still for some time and turn off any source of moving air to get a valid response.

If your sensor provides a change in signal when you move your hand in front of it, it's passed the test and you can move on to the physical mounting section. If you don't see a change in signal, proceed to the troubleshooting section, where you cut those little sensor gremlins off at the pass.

Troubleshooting the motion-detector sensor

If your sensor isn't working, the problem is rarely a defective sensor. However, it is possible for incorrect wiring to damage a sensor; in that case, the sensor has probably become unusable. We all lose a sensor at one time or another. Consider it a battle scar and move on.

The only thing that can usually go wrong with your motion-detector sensor at this point is incorrect wiring. Use the following to troubleshoot problems with the motion detector:

✔ **Incorrect voltage supply:** If you don't have a data sheet, you simply have to get one from somewhere. (This is one situation where you can't fly by the seat of your pants.) Read it and make sure the supply voltage is 5 volts. Also, check the output signal specs to make sure they specify a range of 0 to 5 volts.

✔ **Wrong pin-outs:** This is the big troublemaker. Here's your pin-out checklist:

- Make sure you understand which pin is which. The expansion connector has 40 pins and it's easy to get them confused. Pin numbers 1 and 2 are on one end of the connector, and pins 39 and 40 are on the other. All the even pins are in one row, and all the odd pins are in the other. Identify pin 1 and mark its location on the bottom of the expansion board. Remember that the bottom of the board is the reverse image of the top view.

- Make sure you understand which view of the module the data sheet is showing you by matching the physical sensor to the drawing.

- Follow each wire from the module to the expansion board and check for correct connections.

✔ **Bad solder connections:** Use a good light source to investigate your solder connections. The kind of work light with a spring-loaded arm is the best because you can move the light right where you need it. Make sure each connection looks bright, shiny, and round, and double check for shorts.

✔ **Broken wire:** It's possible, but not likely, that you have a broken wire. Sometimes wires can break inside the insulation, especially wires with solid conductors. It's not the first thing I'd look for, but it's something to check if all else fails.

The assembly: Putting the sensor in place

As with real estate, the location of a motion detector is everything. It's not a good idea to mount the motion sensor on the head perf board because moving the head will cause erroneous readings. (Also, you'll be using that area for something fun in a future project.) I suggest mounting the detector on the front of the head's servo motor. It's a nice flat surface that faces forward, and double-sided tape will stick to it nicely.

I've used several premade PIR sensor modules, and for some reason, the plastic lens on the front is not attached very well. It's easy to pull it off, and I certainly don't want it to fall off during operation and get lost or eaten by the dog. I suggest using a small drop of glue on each side of the lens to make sure it stays on the circuit board. If the lens does fall off, be sure to check out the PIR sensor inside to see the sensor element and related electronics.

Here are the steps you should follow to mount the sensor on the robot:

1. **Turn off the power to the robot.**

2. **Place the sensor on the servo motor's body and make sure the wires can reach.**

3. **Cut a small ¾-x-¾-inch piece of double-sided tape or Velcro to fit the back of the sensor.**

4. **Apply the tape to the sensor first, and then carefully stick the sensor on the motor.**

   After the tape is on there, it may be difficult to remove, so apply it carefully.

Figure 14-10 shows the motion detector being mounted to the robot.

The Brains of the Beast: The Software

Motion detection isn't brain surgery, but you still need to provide your robot with a program to make it happen. As with the other projects in this book, the motion detector has a low-level routine that separates interaction with hardware from the high-level software.

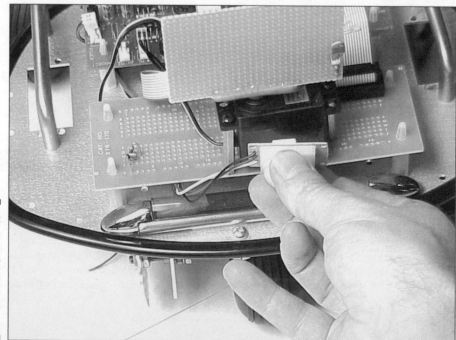

Figure 14-10:
Mounting the motion detector to the head servo motor's body.

It may seem silly to break up this code because only one command is needed to read the motion detector, and it could easily be put inside a high-level routine. Still, modularity is worthy concept, a healthy habit, and you'll be thankful if you ever have to change the hardware.

Low-level code

The first step is to create the code that reads the motion detector. Listing 14-1 shows a routine that can be used by any high-level program that needs access to the motion-detector-sensor hardware. The variable definition is required (see Chapter 6 to brush up on programming), so be sure to place it at the top of your program.

Listing 14-1:   Reading the Motion Detector

```
motiond var bit          'Motion variable

'--------------------------------------------
'getmd
'Reads the PIR motion detector and sets the
'variable motiond.

getmd:
       motiond = in5     'Get detector state
       return            'Done
```

As you can see, the routine needs only one command to read the motion detector. The code uses signal P5 on the Basic Stamp 2 controller. It's possible to use any unused pin; I selected P5 to prevent interference with the requirements of other projects in this book.

High-level code

Oh, you say you'd like to see some kind of motion detection actually happen about now? Well, to make anything happen, you need a simple high-level program to call the getmd routine and display the results.

In Listing 14-2, the red LED is used to indicate the status of the motion detector. Remember to place the variable definition at the top of the program and the getmd routine at the bottom. (You can download the testmd.bs2 program at the companion Web site.)

Listing 14-2: Testing the Motion Detector

```
'------------------------------------------
'testmd.bs2  - Roger Arrick - 5/23/03
'
'This program tests the PIR motion detector
'by reading and turning on the red LED when
'motion is detected.

motiond var bit          'Motion variable

start:  gosub getmd       'Get motion detector
        if motiond = 1 then start1
        high 10           'Red LED off
        goto start        'Loop forever
start1: low 10            'Red LED on
        goto start        'Loop forever
```

The comments to the right of the lines of code describe each step of the program. Examining code like this is a good way to find out more about programming.

Running the testmd.bs2 program involves these steps:

1. **Connect the serial cable to the robot and to the computer.**

   See Chapter 10 for details on this step as well as running the editor.

2. **Turn on power to the robot.**

3. **Open the Stamp editor program.**

4. **Open the testmd.bs2 program in the editor.**

5. **Click the Run button in the editor to transfer the program to the robot.**

After a brief warm-up period (1 to 10 seconds), the sensor should begin responding to movements of your hand by turning on the red LED. The sensor's response is not fast, and each trigger lasts a half-second or more. You'll notice that the sensor is very sensitive to motion nearby and less so with motion that's farther away.

Don't get confused by moving sources of heat such as warm air from a vent or roaming pets. Position the robot to minimize false readings caused by these things.

Troubleshooting

Because you've already tested the hardware with the high-level code, and because the software is so simple, you shouldn't have much trouble making the system work. But as you've probably noticed by now, if something can go wrong it probably will, so if you encounter problems, check this list of possibilities:

✔ Is the download cable attached on both ends?

✔ Have you made any typographical errors in the listing?

✔ Are fresh batteries in the robot?

✔ Is the expansion cable connected on both ends?

✔ Is the power turned on?

Remember, if you get stuck, see the companion Web site for more information or to submit questions.

Real-World Applications

Now that your robot can sense motion, what fun could you get up to? You could put it in front of the refrigerator so it can let you know every time your spouse, who is supposed to be on a diet, goes for the leftovers. Or how about letting it sense how many times your dog goes in and out of the bathroom to drink from the toilet?

These ideas may be silly, but they aren't that far fetched: you'll find that the PIR sensor is great for security and other applications where the robot needs to sense human motion.

A couple of short applications are shown here to spur your imagination. Feel free to use snippets of the following code and ideas to piece together a larger and more complex project.

Doorway greeting

Want to give your guests and housemates a unique greeting when they come through the door? Try this: Your robot can sit and wait for someone to walk in and respond with a programmed action that could include sound, motion, or lights. (Wouldn't it be a hoot to greet guests with a rendition of "If I Knew You Were Coming I'd Have Baked a Cake"?)

If you're interested in tracking your teenagers' comings and goings, you could also have the robot note the time each entry occurred by writing a program that captures the time and stores it in memory. Later you could download the data to your PC, in the same way that you handle temperature data in Chapter 13.

This program makes use of the `getmd` routine and is fairly simple. The program is shown in Listing 14-3 and is well commented to show you what's going on.

Listing 14-3:   Greeting Visitors

```
'------------------------------------------------
'greet.bs2 - Roger Arrick - 5/23/03
'
'This program waits for someone to walk by, then
'greets them with a doorbell-like sound and an LED.

motiond var bit                'Motion variable
waitt var byte                 'Wait time

start:  pause 10000            'Wait for warm-up

greet1: high 10                'Red LED off
        gosub getmd            'Get sensor reading
        if motiond = 1 then greet2
        high 10                'Red LED off
        goto greet1            'Loop forever

greet2: low 10                 'Red LED on
        'Doorbell sound
        freqout 9,600,800      'Alarm sound
        freqout 9,600,650      'Alarm sound

'Wait for motion to stop for 10 seconds
greet3: waitt = 0              'Wait time
        low 10                 'Red LED on
greet4: gosub getmd            'Get sensor reading
        if motiond = 1 then greet3
        high 10                'Red LED off
        pause 500              'wait .5 seconds
        waitt = waitt + 1      'Update counter
        if waitt = 20 then greet1
        goto greet4            'Loop until finished


'------------------------------------------------
'getmd
'Reads the PIR motion detector and sets
'variable motiond.

getmd:
        motiond = in5          'Get detector state
        return                 'Done
```

A simple way to set up this program is to have the robot make a sound whenever an object passes by the detector. However, the result would be annoying because a single object or entity would result in multiple triggers. That's why the code imposes a delay after an object is detected. The code dictates that the field must be clear of motion for a period of 10 seconds before a new event can occur. This is a simplified form of having the software filter the sensor's data, and it works nicely.

After downloading the program, place the robot near a doorway so that people who walk through are detected, as shown in Figure 14-11. Placing the sensor so that it can see through the doorway will cause unwanted triggers when a person gets close but doesn't enter, so place the robot inside the room and behind the door frame (but not in the swinging door's path to avoid robot disasters).

The two tones produced upon triggering sound like a doorbell, but you can easily modify the program to create any series of tones you want — even the sound of a telephone ringing. Adjust the parameters of the `freqout` commands to achieve the results you want.

The first parameter after the `freqout` command determines the pin number connected to the speaker and should always be 9. The second parameter is the duration of the tone (1–65535). The third parameter is the frequency (1–32767). Use the `pause` command between `freqout` commands to make delays between tones.

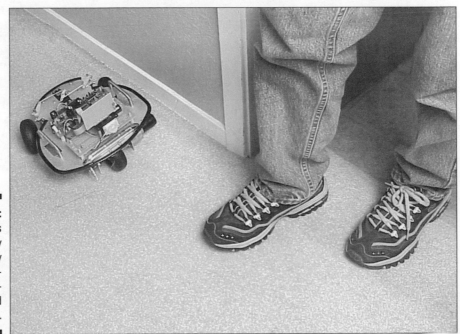

Figure 14-11:
Humans are easily busted by the motion-detector-equipped ARobot.

Pest alarm

Does your dog always get on the couch when you're gone? Of course it does — wouldn't you jump on a nice, soft, couch if your human left you alone? A derivative of the doorway greeting application can be used as a pest alarm to help stop this unacceptable behavior on the part of Midi, Lori, or Bradley.

The code looks about the same as the doorbell greeting, but a different response is included. Instead of a pleasant doorbell sound, this application calls for something more startling to make your dog jump several feet in the air and learn its lesson.

The code creates a randomized sequence of irritating tones that change rapidly. To banish that old hound from the couch forever, simply replace the freqout commands in the greet.bs2 program:

```
'Doorbell sound
freqout 9,600,800   'Alarm sound
freqout 9,600,650   'Alarm sound
```

with this code:

```
'Make crazy randomized sound
i var byte            'Loop counter
r1 var word           'Random #

for i = 1 to 20       'Loop 20 times
  random r1           'Create a random #
  r1 = r1 / 31        'Make # smaller
  freqout 9,70,r1     'Make short random tone
next
```

The new code is sure to make your favorite pest — uh, pet — avoid the area.

Chapter 15

Yakety-Yak: Adding Speech to Your Robot

*I*f your robot could talk, what would it say? How about telling you the status of a sensor, informing you of the distance it's traveled, or maybe just lodging a complaint about hitting an obstacle in its path (ouch!). If your robot could talk, it could give warnings to innocent pets as it navigates through your living room, sound alarms in security applications, or provide robotic chit-chat to amuse a lonely robot builder. Well, robots *can* talk — courtesy of electronic speech technology.

In the past, speech production was difficult and required expensive hardware. As time went on, electronic speech became more commonplace and even found its way into high-end toys and fancy electronic equipment. In the last few years, incredible changes have occurred in the availability of electronic speech technology, which is now so common that even inexpensive toy bears and greeting cards can chatter away. That's because the integrated circuits that make up the guts of speech systems are now cheaper to produce, and more are being manufactured. Producing more of something makes it even cheaper.

In this chapter, I'll get you up to speed on the latest in speech technology, and show you how to jump on the bandwagon by adding speech capability to your robot. You'll find your mechanical friend to be far better company if it can talk back.

Straight Talk about Speech Technology

Humans have been making machines that talk for decades. The first devices were completely mechanical, with a source of vibration to emulate the vocal cords and a resonant cavity to emulate the oral tract.

When vacuum tube electronic devices appeared, engineers designed audio oscillators that could be used to create the sounds that made up the various frequencies that human speech is composed of. Then digital computers came on the scene and were used to save and retrieve sounds by converting them to numbers, which they stored in memory. Initially, computer memory was so expensive that much of the focus of sound technology was on compressing bits of sound so that they would consume less space.

Today, computers are cheap and memory is abundant, so it's easy to create and store sounds. The newer speech technology is focused on attempting to simulate the connection of various sound pieces in a way that produces speech that flows naturally, just like the human voice. The ultimate goal, aside from intelligibility, is to simulate speech so well that humans can't tell the difference between your speech and that of a machine.

Electronic deflation to the rescue

Digital and analog are the dynamic duo when it comes to storing and playing back sound information. *Analog* refers to a device or a signal that constantly varies in strength, such as an audio signal. *Digital* relates to the use of binary digits to code information in patterns using 1s and 0s.

Much of today's speech technology uses digital information to store sounds and analog to play it back. Telephone message machines, voice recorders, and talking business cards are all examples of speech that's been encoded in digital form, saved in memory, and then converted back into analog signals through speakers.

The declining price of memory used to store digital data has played the most important role in enabling our modern-day machines to talk.

Figure 15-1 shows a typical message record/playback chip that can store up to 60 seconds of speech — either a single long message or many shorter ones. The chip requires a few external components, such as a speaker, to make it completely operational.

Toys on the leading edge

Maybe because toy manufacturers are kids at heart, toys are often the first products to use exciting new technology. Although expensive for their time, educational toys were the first mass-produced products that used electronic speech successfully. Nowadays, it's hard to find a toy that doesn't talk, and it's all accomplished with digital recordings of sounds. The figure shows an educational toy that helps kids learn to read.

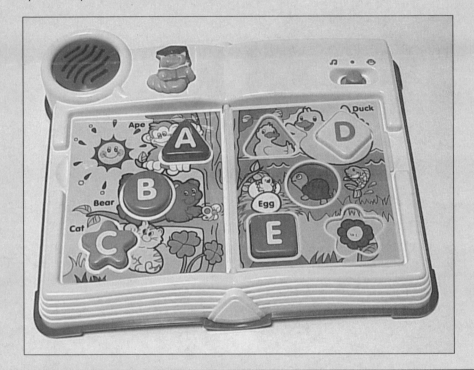

The power of text-to-speech systems

Many speech products available today simply record and play back sound, but another type of technology called *text-to-speech* (TTS) takes text and converts it to a continuous stream of understandable speech. Doing this conversion is speech technology's version of complicated brain surgery: It requires a fast computer and some fancy programming to figure out what pieces of sound to link together to create speech that matches the text.

Punctuation is also taken into account to create changes in pitch such as a rise in the pitch at the end of a question. Often, these text-to-speech systems even allow you to select the type and style of voice used, for example male, female, fast, or slow.

Because TTS systems are much more complex than message recorders, they consume more power and are bigger, not to mention much more expensive. In spite of all that, a TTS would be a great addition to ARobot because it could say virtually any words you send to it, not just a few short sound bites. So, if you have the funds and inspiration, I suggest you look into a TTS system. Figure 15-2 shows the powerful TTS system from RC Systems (www.RCSYS.com).

Figure 15-1:
Today's record/play back chips rely on abundant, cheap internal memory.

Figure 15-2:
Text-to-speech systems can say much more than message recorders.

Building a Speech System

A fella named Ralph Waldo Emerson once said "Speech is power." To get your robot empowered with speech, you need to build yourself a speech system.

Building a speech system is relatively easy if you start with a prebuilt module that includes a speech processor, such as the one shown in Figure 15-3. Most of the remaining work is in wiring and mounting the processor to the robot. After soldering a few wires, drilling a hole, and mounting it, you'll be on your way to having a powerful talking robot.

Getting speech off-the-shelf

Easy is always the first way to go, for my money, and the easiest way to get your robot talking is to buy a ready-made speech product. Many vendors sell small circuit boards that contain all the components necessary to produce speech from prerecorded messages. You may also find more expensive text-to-speech systems, or you can buy individual chips to make your own recorder. (Starting with a chip and making a sound module by adding more parts is not something I'm interested in doing just to save a couple of bucks, but I don't want to discourage you from doing that.)

Figure 15-3:
ARobot's
new speech
processor.

For this project, you'll use a prebuilt message recorder module. Your only tasks will be to mount it, interface it to the robot, and write some software.

Several companies offer speech products designed to be embedded in machines. These products are usually intended for speaking vending machines or talking toys, but nothing says you can't use them on your robot. When selecting a speech product, it's important that it have certain qualities. It should be

✔ As small as possible

✔ Inexpensive

✔ Consume little power

✔ Capable of interfacing to ARobot's controller

✔ Simple to program

A Web search for *speech recorder, message recorder,* and similar phrases will return a large number of links to commercial products. You may want to add the word *kit* or *robot* to the search phrase because you're looking for a do-it-yourself type of product that's designed to be added by the user to a larger system such as a robot.

Here are some speech products that are currently available:

Product:	Sound Module AppMod
Company:	Parallax Inc
Web site:	www.parallaxinc.com
Price:	$80
Assembled:	Yes

Description: Small (2½-x-2½-inch) circuit board. Stores and records up to 60 seconds of sound. Includes small amplifier and speaker. Easy programming through serial communications to access multiple messages. Microphone included for recording.

Product:	K146 Message Recorder
Company:	Kits R Us
Web site:	www.kitsrus.com
Price	$20
Assembled:	No

Description: Small (2-x-2½-inch) circuit board. Stores a single message up to 40 seconds but can be modified for multiple messages. (Requires several port pins to access multiple messages.) Includes a microphone but the user must supply a small speaker.

Product: V8600A Text to Speech Synthesizer

Company: RC Systems

Web site: www.rcsys.com

Price: $130

Assembled: Yes

Description: Circuit board (about 3 inches by 4 inches). Extremely powerful system that can and create good-quality speech in various voices from text. Can be controlled with a single serial line. Low power consumption. User must supply a speaker.

Out of these possibilities, we'll use the Sound Module from Parallax. It can produce multiple messages and requires only a single port pin from the Basic Stamp 2. It's also completely assembled and includes a speaker, so you can focus on getting the robot talking instead of soldering components and searching for parts. Figure 15-4 shows the module, hardware, and documentation.

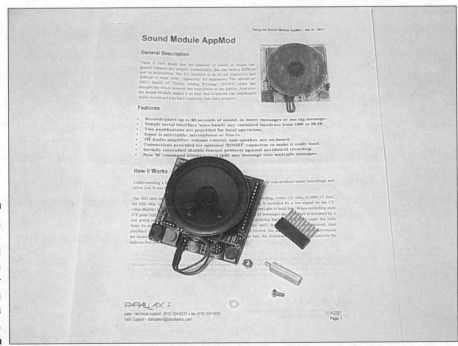

Figure 15-4:
A fully-assembled message recorder module saves us some time.

Collecting the parts

The parts list for this project is a brief one because the sound module includes the hardware and a speaker (see Table 15-1). After scrounging some wire, you'll have all the parts you need. Any wire from awg 20–26 is fine, but stay away from solid conductors because they're not flexible enough and tend to break.

Table 15-1	Speech System Parts List		
Part	*Vendor*	*Quantity*	*Approximate Price Each*
Sound Module AppMod	Parallax	1	$80.00
6-inch wires, awg 24	Anywhere	3	$0

Wiring

Time to get wired for sound! The sound module requires only three connections. I prefer using three different color wires if possible:

- +5 volts (red)
- Ground (black)
- Communications signal (a third color such as green)

Color-coding helps you with troubleshooting efforts down the road.

Follow these steps to connect wires to the sound module:

1. **Cut the three wires to 6-inch lengths and strip each end about ¹⁄₁₆ inch.**

2. **Tin each end with a dab of solder.**

   Tinning allows a connection to be made without having to add solder with your nonexistent third hand.

3. **Locate the three pins (+5, ground, communications signal).**

   The bottom of the sound module has a 20-pin connector, which is described in the documentation. Pin numbers are not identified because the module is designed to plug into another product. You'll have to rely on Figure 15-5 and my description. Hold the module with the bottom of the board facing you and the speaker facing away. You'll see a single mounting hole near the bottom center. Orient the module so that the connector pins are across the top. There are two rows of ten pins each. The pin on the top left is +5 volts.

Figure 15-5:
Double-check the wire and pin placement before soldering. Check again when you're finished.

4. **Solder one end of the red wire to the area where the connector pin meets the board.**

   See Figure 15-5 for help.

5. **Go to the pin on the far-right top and solder the black wire to it.**

6. **Find the fourth pin from the left on the bottom row, and solder a third color wire to it.**

   Again, Figure 15-5 shows you the correct pin for this connection.

Now connect the other ends of the three wires to the bottom of ARobot's expansion board by following these steps:

1. **Solder the red wire to the +5 volt rail.**

2. **Solder the black wire to the ground rail.**

3. **Solder the third color wire to pin 12 of the expansion connector.**

Figure 15-6 shows the details of connecting these wires to the expansion board.

It's important not to get the wires mixed up. Otherwise, you could damage the sound module.

Figure 15-6:
Solder the wires to the bottom of the expansion board.

Mounting the speech module

It's time to introduce your robot to its new vocal cords (that is, the speech module). The speech module has one hole intended for mounting it on a surface, and the package includes a spacer and screws. By using the spacer, which is almost 1 inch tall, you can mount the module on the expansion board without consuming valuable circuit space because the spacer raises the module up above the board.

Follow these steps to mount the speech module:

1. **Attach the spacer to the module by using a screw on the top and the spacer on the bottom.**

   It takes time to get the screw into the hole because the speaker is in the way.

2. **Drill a ⅛-inch hole on the end of the expansion board, as shown in Figure 15-7.**

3. **Attach the module to the expansion board using another screw.**

   Make sure the wires are routed in a clean way.

Figure 15-7:
The spacer
raises the
module
above the
circuitry
on the
expansion
board.

Testing

Want to see whether you have the right wires in the right places? Here's how: Apply power to the module by turning on ARobot's power switch. The LED on the sound module should flash red and green, indicating that it's in *protected mode,* in which memory can not be overwritten. That's a good sign because it tells you that you wired the power correctly.

You can also test the module manually by using the pushbuttons on the module, but you'll have to do a few things to allow that to happen. First, the onboard microphone must be enabled by ensuring that jumper X2 is on pins 1 and 2. The unit is normally delivered in this condition. Next, you must send a command to the module using a short program that enables recording. Listing 15-1 is the renable.bs2 program, which is available also from the Web site at www.robotics.com/rbfd. Refer to the user guide that accompanies the module for the details of operation.

Listing 15-1: Enabling Message Recording

```
'-------------------------------------------
'renable.bs2 - Roger Arrick - 5/23/03
'
'This program enables recording on the sound module.
'Make sure jumper X2 is on pins 1 and 2 to enable the
'microphone before recording your message.

    pause 500              'Start delay
    serout 7,188+$8000,["!SMON"]  'Enable Recording
    end                   'All finished
```

There's no need to understand the details of the program at this point. After you download this program into ARobot, it executes automatically.

This documentation lists all the commands that can be sent to the sound module by the Basic Stamp 2 to control recording, playback, and a few other utility features. I list all the important ones in the "The sound module command reference" section, later in this chapter.

After the program has run, you'll notice that the LED on the sound module has turned to green, indicating that the unit is not in protected mode. You can now use the pushbuttons on the sound module to record a message:

1. **Push the black button until the LED turns red (record mode).**

2. **Push and hold the blue button to start recording.**

3. **Speak your message into the microphone.**

4. **Release the blue button to stop recording.**

    Your message is now saved in digital memory inside the sound module.

To play back this message, simply push the black button until the LED is green (play mode), and then press the blue button to play the message back.

Experiment with different voice levels and distances from the microphone to achieve the best recording results.

Troubleshooting

Things going haywire? If you're having trouble with the speech module, the main thing likely to have gone wrong is the wiring. It's important that you solder the wires to the correct pins and make sure that there are no shorts. You can destroy a sound module by miswiring it, which is bad for your budget and leaves your robot speechless.

Look over your wiring job to make sure you wired it correctly:

✔ If the LED on the sound module doesn't turn on when you power up ARobot, power, ground, or both are not connected correctly.

✔ If the LED flashes red and green, but running the program in Listing 15-1 doesn't make the LED stay green, the communications wire is not connected correctly. The most likely error is that you counted the pins on the expansion connector incorrectly, so check that.

Another reason that the program might not work is a typographical error. For example, the following code, which appears in Listing 15-1, contains the number 0, not the letter O:

```
["!SMON"]
```

Also, the commands are case sensitive, so make sure that all the letters are uppercase and don't include any extra spaces.

Creating Sound Software

Several commands can be sent to the sound module to get your robot talking. The documentation provided with the sound module describes each command, but you will need to make a change to the parameters because I've used port P7 on the Basic Stamp instead of port P4 as shown in the documentation. (I did this because P4 is already being used by ARobot's light sensor.)

The commands that you send to the sound module are simple and consume only one line of code. Intelligent modules that require short commands (such as this speech module) are effective at offloading work. In this way, the Basic Stamp 2 isn't consumed with the details of controlling complicated hardware devices, and it has more space in memory for even more elaborate programs. This section explains how to send the commands to ARobot and provides a brief command reference to help you explore sound possibilities.

Communicating with the sound module

As with every other project in this book, installing all the nuts and bolts and thingamabobs is only the first part of what you need to do to get your robot ready for action. Now you have to send a program to the robot containing commands that tell it what to do.

You send each command to the sound module using the `serout` instruction (within a program), which sends data using a single port pin in a serial fashion. Here is a typical command:

```
serout 7,188+$8000,["!SMOP",0]
```

The line begins with the `serout` instruction followed by a series of parameters. The first parameter, `7`, indicates port pin P7, which is used to transfer the data. The second parameter, `188+$8000`, indicates the speed and other specifications for the data transfer. The data to be sent to the sound module is `["!SMOP",0]`.

The sound module has its own special command structure. All commands begin with `!`, which tells the module that a command follows. The `SM0` portion tells sound module zero that the command is intended for it and no other. In this example, this is followed by a `P` to indicate play mode. The `0` tells the sound module which message address within the sound module to play.

This elaborate command structure allows several devices to be connected to and share a single port pin. Each device listens for data and responds to only the commands intended for it.

It's not important that you remember each command because you'll be using only a few. In addition, the commands are included in example programs, so you can cut-and-paste them as you need them.

The sound module command reference

This section contains a list of commands you can use to control the sound module. After you record messages, most of your programs will simply use the playback command, so don't get overwhelmed with all the options on this list. Just choose ones you'd like to try and have fun!

Before introducing the sound module commands, I'd like to mention a bit about message addresses. The sound module has a single block of memory, which can contain one long message or several individual messages. You access messages for playback or recording by specifying the starting address within this block of memory.

Enable recording

You can't record without first sending the enable recording command to the module.

Example:

```
serout 7,188+$8000,["!SMON"] 'Enable recording
```

Record message command

The record message command begins recording at the specified address. Each number in the address represents 0.4 seconds. The first address location is 0. Location 10, for example, is 10 x 0.4 = 4 seconds.

After recording has begun, you can stop it by sending an end command (E), as described in the next section.

Examples:

```
serout 7,188+$8000,["!SMOR",50] 'Record from 20-second mark
serout 7,188+$8000,["!SMOR",25] 'Record from 10-second mark
serout 7,188+$8000,["!SMOR",0]  'Record from beginning
```

End recording command

All good things must end. The end recording command simply stops the recording in progress and sets the end-of-message marker.

Example:

```
serout 7,188+$8000,["!SMOE"] 'Stop recording
```

Playback message command

The playback message command simply plays a message at the specified address. This is the command you'll be using most often in your robot programs.

The trick to this command is the last parameter, which specifies which message (at which memory address on the sound module) to play. Each integer number represents a 0.4-second time slot. So, for example, if the last parameter is 75, the command plays the message starting at the 30-second address mark (because 75 x 0.4 seconds = 30 seconds). Playback stops when an end-of-message is reached, an I command is given, or the end of memory is reached. The maximum size of a single message is 60 seconds.

Examples:

```
serout 7,188+$8000,["!SMOP",75] 'Play from 30-second mark
serout 7,188+$8000,["!SMOP",10] 'Play from 4-second mark
serout 7,188+$8000,["!SMOP",0]  'Play from beginning
```

Initialize command

The initialize command resets the module. It's useful if the module stops responding due to overflow or other errors and quits working. Messages in memory are not lost.

Example:

```
serout 7,188+$8000,["!SMOI"] 'Initialize the module
```

Recording sound

If you're thinking that you'll have to write a program just to record each message, you're right. But you'll be happy to hear that I've written a program to simplify the entire process of filling the sound module with custom messages.

This section outlines the soundr.bs2 program, which hides the complexities of recording a message, setting the end-of-message marker, and so on. The program is simple to operate and automates the process of creating a bank of useful messages that you can then play back in programs that you write for your robot.

Sound recording utility operation

The soundr.bs2 program can record up to thirty 2-second messages. That 2 seconds is enough time to say a word or two, which is plenty for most robot applications. After downloading the program into ARobot, it waits for you to press the SW1 or SW2 button.

The SW1 button controls the recording process. Press it once and say a word into the microphone. You'll notice that the red LED turns on during recording. When recording has stopped, you'll hear two beeps and the LED returns to green. Press SW1 again to record the second message, and so on. After recording the messages, press SW2 to play them back one at a time.

Listing 15-2 shows the soundr.bs2 program. You can save yourself some time by downloading the pretested program from the Web site at www.robotics. com/rbfd.

Listing 15-2:   Recording 30 Messages

```
'-------------------------------------------
'soundr.bs2  - Record 30 messages
'
'This program allows you to record 30
'messages of 2 seconds each.
'Make sure jumper X2 is on pins 1 and 2 to enable the
'microphone before recording your message.
```

```
'To operate, press SW1 to begin recording.
'After 2 seconds, the robot will beep.
'Then press SW1 again to begin recording the next message.
'Continue until all messages are recorded (30 total).
'You can then press SW2 to play each message.

        r var byte           'Record message counter
        p var byte           'Play message counter

Start:  r = 0                'Init message counters
        p = 0
        pause 500            'Start delay
        serout 7,188+$8000,["!SMON"] 'Enable recording

sw:     if in14=0 then rec   'If SW1 pressed then record
        if in15=0 then pla   'If SW2 pressed then play
        goto sw              'Loop until button pushed

rec:    p = 0                'Restart play counter
        serout 7,188+$8000,["!SMOR",(r*5)] 'Record 2 sec
        pause 1950           'Wait till recording done.
        serout 7,84+$8000,["!SMOE"]  'End recording
        pause 50             'cover up the beep :)
        freqout 9,100,2000   'beep
        if r = 29 then rec1  'If no more space then done
r = r + 1            'Point to next message
        goto sw              'Message loop
rec1:   r = 0                'Overflow to beginning
        pause 50             'Delay between beeps
        freqout 9,100,2000   'Second beep
        goto sw              'Done with recording

pla:    r = 0                'Restart record counter
        serout 7,188+$8000,["!SMOP",(p*5)] 'Play message
        pause 2000           'Wait till played
        if p = 29 then pla1  'Last message?
        p = p + 1            'Point to next message
        goto sw              'Done with playing
pla1:   p = 0                'Overflow to beginning
        goto sw              'Done with playing
```

Playing messages

Okay, you've been patient. It's time to reward yourself by telling you how to get your robot to speak to you. You do that by sending the command to play a message.

The messages are numbered 0 through 29 (for a total of 30). Here's a command to play the third message, which is numbered 2 because we start from 0:

```
serout 7,188+$8000,["!SMOP",(2*5)]  'Play message
```

To play a different message, simply change the 2 to another number between 0 and 29. It's that simple.

Building Your Robot's Vocabulary

As you start to figure out exactly how you'd like your robot to use its new power of speech in real-world applications, it makes sense to create a standardized list of words and phrases that make up your robot's vocabulary. After careful thought about what information you would need from your robot, I came up with the following list. Message numbers are on the left.

0	No	6	Stop!	12	Light level	17	Low
1	Yes	7	Starting	13	Moving	18	Medium
2	Hello	8	Done	14	Motion detector	19	High
3	Ouch!	9	On			20	OK
4	Warning!	10	Off	15	Forward	21	Left
5	ARobot	11	Whisker	16	Reverse	22	Center
						23	Right

That leaves room for 6 more messages for special purposes such as calling your name or notifying you of an error. Feel free to rework this list to include the numbers 0 through 9 for saying the results of sensor measurements, specific warnings, or specific actions. You can even record phrases from the radio or movies for great special effects! When you've finished compiling your list, use the soundr.bs2 program to record it.

Be sure to make a list of these on a note card and post it near your computer to aid in future programming.

Putting Speech to Work

Using speech in your robot programs is easy because it requires only a simple command to produce a message. But why not get creative? Don't forget that you can concatenate, or join, sounds such as "light level high" or "motion detector on." In addition to hearing your robot telling you what it's doing, you can have fun by having it greet you by name, tell the dog to scat, or wish you a happy birthday.

Here are some real-world applications that use speech in your robot.

Debugging

The first place you'll find speech useful is in debugging your programs. Because your ARobot doesn't have a display (yet!), you can use sounds to tell you what stage the program is at to help find errors and logic problems.

Don't forget that a sound takes a finite amount of time to play and the program may be continuing with other instructions, so put a `pause 2000` after a 2-second message to synchronize the speech with the program.

Navigation

Having your robot speak as it navigates around your house is fun. When your robot bumps into a coffee table leg and yells "ouch!" everyone will laugh. And hearing your robot tell you the condition of sensors, such as "light level low" (or better yet, "Hey, it's dark in here!") makes your robot seem alive.

Safety

Better safe than sorry definitely applies to your robot's daily antics. Having your robot announce its intentions, such as "starting" or "moving forward," could prevent an accident or two. Believe me, getting to an emergency room on a busy Saturday when you've been tackled by a suddenly moving robot is nearly impossible (not to mention how embarrassed you'll be trying to explain exactly how you tripped over a mechanical friend and fell over the coffee table).

Security

Who needs a guard dog when you have a robot in the house? Security applications are commonplace in robotics and there are some interesting ways that speech can enhance a robot's performance and utility. Having a robot yell "stop!" or "warning!" might convince an intruder that humans are nearby.

Data collection

Data collection is an often-undervalued application, but it's an excellent use of your mechanical friend. Speech can play an important role in transmitting that data to you. For example, suppose you have a program that measures the temperature every ten minutes throughout the day and records it in memory. The operator could come along with a notepad and pencil, press a button, and record the results as the robot repeats them. Effective and fun!

Making music

It's possible to record short music segments or make your own by recording notes from a keyboard, and then have your robot play the notes in a certain sequence to create a tune.

The recording process uses the soundr.bs2 program modified for 0.4-second message lengths instead of 2 seconds. This gives you 150 different messages stored in the sound module, and because each octave is 12 notes, you'll have more than enough room to record a piece of music. Record each note using a keyboard instrument, piano, or any other instrument for that matter. Start the note before pressing the record button so it will consume the entire message time.

The main limitation with this method is the 0.4-second message length of the sound module. It would be nice to make shorter messages, but it's just not possible due to the limitations of the recorder chip on the sound module. But you can make longer messages by making a batch that is twice as long and a batch that is four times as long. The sound module's memory will hold a couple of octaves worth of notes in this case.

Talking about Robots

Now that your robot can speak, you'll probably notice people responding to it differently. Something about hearing a robot make sounds adds a touch of life to the little mechanical guy. It gives us a feeling that the machine is somehow conscious because it shares our ability to communicate.

Well, you know it's just a circuit and a series of instructions, but you'll enjoy your little chats with it anyway. Besides, you have a new project on your hands: making your robot be quiet!

Chapter 16

I See You

*T*hey say a picture is worth a thousand words, so enlisting your robot buddy to go out and feed you pictures with an onboard video camera must be worth . . . at least a chapter.

Now you should know that most robots don't have anything approaching real vision, which would involve not only capturing images but processing image information. Even for today's high-speed computers with hundreds of megabytes of memory and giant hard drives, interpreting visual data is like mega-high-end brain surgery. Although these vision systems are available, they are very expensive, complicated to train and configure, and normally confined to industrial robots that do things such as putting parts in a circuit board or positioning an automatic screwdriver.

Instead of a giant, complex system that no one can afford or understand, what you'll discover in this chapter is a simple camera system that feeds video back to you as your robot roams about the world. The hardware required to capture and transmit images back to you, the robot operator, is inexpensive, doesn't require much power, and is easy to implement. And if you get creative, you might find that getting a robot to send you pictures of what it sees can have some interesting uses.

All about Video Systems

One big part of adding a video system to your robot is the video system itself. In this section, you'll get a quick primer on video cameras. First you get up to speed on the current state of video systems. I fill you in on video camera technology, and then deal with monitors and other system options such as audio and wireless possibilities. Finally, I give you a few purchasing tips.

Some general guidelines

Your goal is to simply capture video from a robot-mounted camera (something like the one shown in Figure 16-1) and send it back to the operator. So you're free to select almost any equipment that will do the job.

However, a few things are important to consider. The camera

- Must be small enough to fit on the robot's head perf board
- Must be capable of operating on the low voltage that's available to the robot (12 volts)
- Must consume a minimal amount of current to preserve battery life (less than 200ma)

Keeping those considerations in mind, let's delve into the wonderful world of video system components!

Cameras

Walk down the video camera aisle of any electronics superstore today and you'll soon see that you have many options for cameras. The various models all have different physical characteristics and electrical specifications. Some of those characteristics matter when it comes to using the camera on your robot and some don't, so use the following sections to guide you in your purchase. Then begin reading those data sheets from suppliers.

CMOS ... CCD ... ?

Before semiconductor cameras were available, cameras were made from vacuum tubes that were able to pick up photons and convert them to electrical impulses. These cameras were as big as a small lawnmower and consumed a ton of current. Nowadays, you'll be glad to hear that tubes are ancient history, and small, low-power digital chips are used to produce accurate images at a fraction of the price.

The least expensive type of camera has a CMOS (complementary metal oxide semiconductor) sensing element. This kind of camera is simple to construct but produces more noise than other camera types. CMOS cameras do not produce as good an image when the light level is low, but they're fast at grabbing an image.

CCD (charge-coupled device) cameras are currently the best mass-produced imaging devices. CCDs can also detect objects in very low light levels. Noise and image quality are very good with these types of cameras, but CCDs cost more than their CMOS brethren and are somewhat slower in operation.

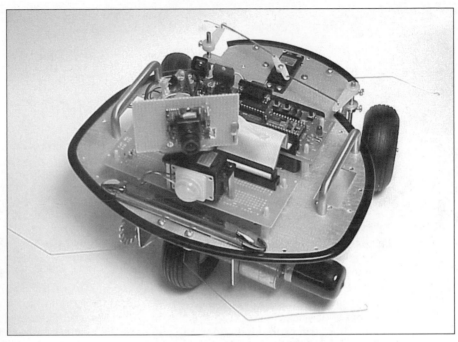

Figure 16-1:
ARobot
sporting a
new camera
system.

Color or black and white

Both CMOS and CCD cameras are available in color and monochrome (black-and-white) models. As you might expect, the color models cost more — often more than double. In a few years, color cameras may be as cheap as monochrome (if anybody even makes monochrome anymore), and then, what the heck, you might as well buy color! Note too that the color models sometimes consume more current.

Which model to buy depends on what you want to accomplish. If you're just watching a warehouse for intruders, monochrome is fine. If you're inspecting a product for flaws, you might need color to make out the details.

Very specific specifications

You know those lists of features that every electronics manufacturer creates for products from high-end stereo equipment to video cameras? Lists that have abbreviations for things you can't begin to figure out? Well, like it or not, studying these lists is the nuts and bolts of camera selection. There's no way around it: You have to sift through a bunch of specifications and try to make a decision about what features are important to you. But I'll help you to decipher those camera specifications.

The _resolution_ of the camera is listed in pixels or lines. This basically tells you how much information the camera will collect for each image and how finely detailed the image will be. A low-end camera is somewhere around 300 x 300 pixels, which is probably fine for most robot purposes.

The *light sensitivity* specification is usually listed in *lux*. This specification is important if you plan on using the system in darker environments. A small number such as .2 lux produces a better image in the dark than a unit with a rating of 10 lux. Typical values range from .1 lux to 10 lux; cheaper cameras have the larger lux value.

Other parameters you might want to consider include signal-to-noise ratio and humidity ratings. Frankly, for our purposes, they aren't that important (unless you want to send your robot into a rain forest to gather images in a high humidity setting).

Size matters

The physical form of the camera is important for our project because the unit must fit nicely on the head perf board. A 4-inch camera that weighs 12 ounces is not going to work. Instead, look for a small board camera, about 1 inch square, that's mounted directly to an exposed circuit board. This type of camera is designed to be installed inside a product or behind a user-supplied panel and is perfect for robot projects due to its size and lightweight construction. There will probably be mounting holes in the corners to accept small screws, which will make it easier for you to mount the camera to the robot.

There are also cameras shaped like a tube of lipstick or in their own plastic case, as shown in Figure 16-2. If you use a lipstick, or bullet, camera, you'll have to use special hardware to mount it to the head perf board. Some small board cameras have a plastic enclosure about 1½ inches square that will work just fine. If the unit comes with a bracket, you might be able to use it to mount the unit to the robot or you might have to remove the bracket, depending on its size and placement. Because of the bulk and extra weight that cameras with enclosures have, I suggest you stick with a board camera if possible.

Lenses

All cameras must have a lens that focuses an image onto the sensor. Most small cameras have a built-in, fixed-focus lens, which means the focus can't be changed. Others have a housing that accepts various sizes of lenses.

The lens focal length determines how large an image will appear. The greater the focal length, the more the lens becomes telephoto (makes distant objects appear closer). A typical lens on an inexpensive small camera is about 4mm, which is fine for our application.

Illumination

Just as you are likely to bump into things in the dark, cameras don't do very well in low light conditions — especially cameras with CMOS sensors. Because semiconductor cameras are sensitive to infrared light, some come with built-in infrared LEDs (light emitting diodes) that produce light that can't be perceived by humans but is useful at improving picture quality.

Figure 16-2:
Cameras are now available with exposed circuitry and with enclosures.

Infrared LED illumination systems are also available as separate modules to help any camera capture better images in the dark.

Try this interesting experiment: Point your TV remote control at your video camera and notice the flashes on your monitor. Your eye can't see the infrared light, but your camera can.

Interfaces and cables

The interface determines how the image data will be sent to the monitor for viewing. The camera, the transmission components (the cable or the transmitter/receiver), and the monitor should all use the same interface.

Most small cameras produce an NTSC (National Television System Committee) signal, which is known as composite video and is also referred to as the RS-170A standard. Other interfaces such as PAL and SECAM (which are used outside the United States) and S-video are not normally used in small camera systems like the one you're building.

Most small cameras use common RCA-style connectors for the video signal. Cables with RCA connectors and individual connectors for building your own cables are available at most electronic stores and through mail order.

Power requirements

Power requirements and consumption are important items to check. ARobot can supply either +5 volts or +12 volts to power the camera. Most cameras operate on 7 to 12 volts, but make sure that the camera you select doesn't require a precise, regulated supply.

Most small cameras will require 100ma to 200ma of current to keep them operating. A large value will drain the batteries too quickly, so look for one with the smallest value.

In a nutshell

In summary, make sure that all the video components use the same interface (usually NTSC), make sure the camera will fit on the robot, and make sure the camera will run on +5 volts or +12 volts unregulated.

Monitors

The term *monitor* refers to a display that accepts video signals from a camera, computer, or other device. Most televisions produced now have video inputs, so they qualify to be called monitors. Display technology is changing rapidly; Figure 16-3 shows the two types that are the highlight of this discussion. I'll skim the surface of display technology to get you up to speed pronto.

CRT

We've all seen a CRT, or cathode ray tube. They're heavy, power hungry, ready to explode, and usually have a display surface that isn't very flat. CRT technology has been around since the late 1800s and has served its purpose well, but it's on its way out, as you might have guessed from your last visit to a home electronics or computer store.

CRTs work by shooting a beam of electrons through a vacuum onto a phosphorescent coating, which glows when hit. The beam is directed with electromagnets, which are controlled by the circuitry that receives the video signal. The result is a fairly good image that we've all become accustomed to.

There's nothing wrong with using a CRT in this application. They are cheap, come in a variety of sizes, and display a fine image.

LCD

The latest display technology is the flat, lightweight, LCD (liquid crystal display), which works by applying an electric current to a liquid crystal solution sandwiched between two sheets of polarized glass. The result is a low-power display surface. Various technologies are used, and every year new models are introduced that offer improved image quality and reduced pricing.

Many types of LCD technologies are available and there's no need to explore each here. Simply put, when you want something small and low power, you'll probably want an LCD. The features of an LCD are especially important in mobile applications.

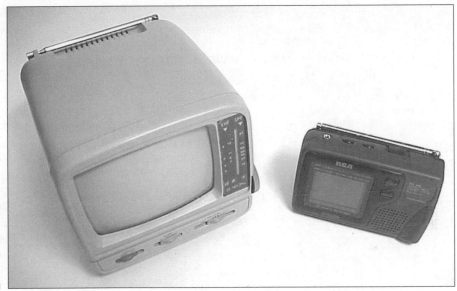

Figure 16-3:
The dying
CRT versus
the LCD.

Making the decision

When you choose a monitor, you'll be faced with sifting through specifications and weighing the cost-and-feature ratio. It's like purchasing a computer or a stereo. Just make sure that the unit has an interface that's compatible with your camera and that the resolution is equal to or better than your camera.

Most LCDs and most CRTs with a composite video input will work fine for this application. Your decision will end up being like many others — how much money do you want to spend?

Tethered or wireless?

The question of whether to make your video system wired (tethered) or wireless usually comes down to money. Most people would like to have a wireless system, but it adds significantly to the cost. Luckily, wireless systems are becoming more commonplace and more affordable. Still, a tethered system has advantages, as well. In this section, I discuss both.

Tethering issues

Tethering is a good solution for remotely controlling or gathering information such as video and audio from a robot. It works well for short distances and usually provides a better signal than a wireless system. Tethered systems are much less expensive than their wireless brothers, but the goal is the same: communicate information between the operator and the robot.

What's an RF modulator?

You may have seen small metal boxes called RF modulators, like the one in the figure, but never understood what they actually did, beyond their use as an interface between video games and other equipment and your TV.

Years ago, most TVs were made to accept transmissions from TV stations only. When video games, video cameras, and VCRs came out, they used composite video and audio signals (usually on RCA connectors) for improved signal quality.

RF modulators take a composite video (and audio) signal and transmit the information as if it was coming from a TV station. The user usually had the choice of transmitting on channel 3 or 4 to avoid interference from a local station. The output of the RF modulator connected to the TV's antenna connector, and there was often a switch box to select between the antenna and the output of the modulator.

Gradually, having composite video and audio inputs on TVs became standard, and RF modulators were discarded. We enjoy much better signals because of this evolution. The progress continues, and now many systems output even higher quality signals such as S-video or RGB. Some new TVs have these interfaces as well as the older composite video interface. Those that don't will require an adapter, or you could just live with the older interface.

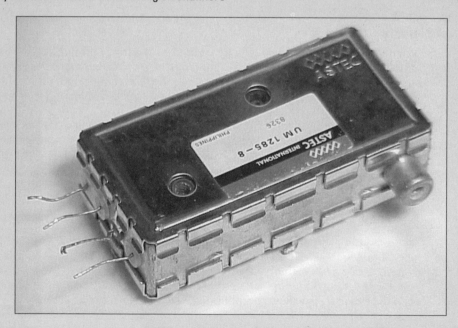

Tethering is accomplished by simply dragging one or more wires behind the robot that connect to other equipment such as a controller or a video monitor. Usually the tether is a single cable with many wires inside to carry video, audio, control signals, and sometimes power. Each of these wires may be different due to the type of signal it carries.

Tethering, however, is not without problems. The cable can drag on the robot and limit its maneuverability. To minimize these effects, make sure the cable is smooth and flexible. The robot builder must also attach the cable to the robot in a way that prevents the cable from pulling loose.

In spite of the disadvantages, tethering is a common method for the remote control of a hobby robot because it is inexpensive and easy to implement.

Losing the wires

You've probably played with a wireless, remote-controlled car, plane, or boat. They're great fun because you have freedom of movement. There's something exciting about pushing a small button and seeing a machine move.

Today, it's possible to buy off-the-shelf wireless products that can communicate data, video, and audio. Commercial products are inexpensive and have limited range; industrial products are more expensive but can have a range of many miles or more with a faster and more reliable data rate.

Many newer commercial video and audio wireless links transmit on special high frequencies and require a matched transmitter and receiver set to operate. Other models are designed to transmit on ordinary television frequencies, which allows you to use a standard TV as the receiver.

Transmitters and receivers typically use RCA-style connectors for both the video and the audio (if they handle both), as shown in Figure 16-4. This allows convenient connection to standard camcorders and VCRs.

Transmitters that use the non-TV frequencies may use add-on amplifiers that extend the range. A special antenna may also be used to extend the range or to improve quality.

To hear or not to hear

Many cameras have built-in microphones to capture audio right along with the video. You'll have to pay a little more but the benefits might be worth it if you need sound for your application.

Many monitors have built-in amplifiers and speakers, which eliminate the need to purchase separate components. Small, amplified speaker systems are available at computer stores and are inexpensive.

If you're using a wireless system, make sure that it's capable of handling the audio. You'll need only one channel, but many systems may be able to handle two channels for stereo.

Figure 16-4:
Wireless
systems
can make
your robot
feel free.

If your robot will be tethered, cabling is also an issue if you're capturing audio. You can use a small coax cable to transmit the audio and keep it noise-free, but single-pair, small-gauge wire also works just fine for most applications.

Monitors and wireless equipment usually use RCA-style connectors for audio, just like video. You can purchase cables with RCA connectors or build your own with individual connectors and cable.

Where to buy?

I'd like to be able to point you to one particular source for all the items needed to make your wireless system work, but you have an abundance of worthy suppliers, each with its own product range, at your disposal. You have so many options and so many manufacturers and the offerings change so frequently that I must insist that you do some Web surfing and choose the best product for your budget and requirements. Just remember that the basic goal is to capture video (and audio if you want) from the robot and get it to the operator — either by wires or by using a wireless system.

For my robot, I chose a small, color board camera and a 2.4 GHz wireless transmitter/receiver from one source, and a small LCD TV with video inputs from another. Here are some suppliers of such equipment:

Supercircuits
www.supercircuits.com
Cameras and wireless systems, monitors, and accessories

Polaris USA
www.polarisusa.com
Cameras and wireless systems, monitors, and accessories

Jameco Electronics
www.jameco.com
Electronic parts, board cameras, and wireless systems

Radio Shack
www.radioshack.com
LCD and CRT monitors, cables, and adapters

Pretesting

After purchasing your camera, wireless transmitter/receiver or cables, and the monitor (LCD or CRT), I suggest that you take some time to test the system before mounting it to the robot. It's always nice to know that the components function before investing time and effort, and it will also give you a chance to check the performance.

One issue you'll have to resolve to test the system is power. Your camera will probably require 12 volts DC, and so might your wireless system. You can supply this power with a wall-transformer-style power source, a bench-top supply, or batteries. Some minor wiring may be necessary depending on your components. Figure 16-5 shows the components of my system being tested.

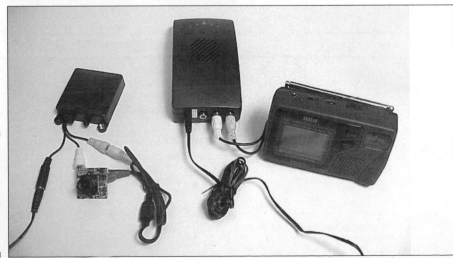

Figure 16-5: The video system should be tested before installation.

Carpenters have a saying: measure twice, cut once. These words of wisdom can be applied to many things in life, including robot building. So, before you apply power to your shiny new video components, double-check the supply voltage, polarity, and wiring. Also, make sure that nothing metal is touching the components and that the components are not touching each other.

Mounting a Board Camera

Mounting of the camera is a separate issue from how the video information will be transmitted back to the operator (tethered or wireless). The camera must be mounted so it won't fall off, and so the cables can be managed. I'll be mounting the camera on ARobot's head perf board so it can be positioned.

Collecting components

I've selected a small color camera with an exposed circuit board and integral lens. The camera module is about 1.3 inches square and will fit nicely on the head perf board. A small connector is located on the back to carry video and power signals, and a short cable is included. The camera, which is shown in Figure 16-6, is small, lightweight, and consumes little power — important features for a mobile robot. Most cameras like this will come with a small but handy data sheet describing the vital details needed for usage.

You'll also need screws, washers, and nuts that match the camera's mounting holes. Small screws are sold in boxes of 100 each, so you'll have to pay more than you might expect. Put the excess in your parts bin for the next project. Other than that, this project requires few components, as shown in Table 16-1.

Table 16-1	Camera System Parts List		
Part	*Vendor*	*Quantity*	*Approximate Price Each*
Color micro board camera PC103XS	SuperCircuits	1	$80.00
Wireless video system MVL10	SuperCircuits	1	$110.00
LCD TV	Radio Shack	1	$80.00
2.1mm DC Connector 1710-2110	Mouser	2	$1.00
#2-56 x 1/2 screw 5721-256-1/2	Mouser	Box	$5.00
#2-56 hex nut 5721-256	Mouser	Box	$3.00
#2 nylon washer 561-D232	Mouser	Box	$2.00

Figure 16-6:
My small board camera with lens, cable, and mounting hardware.

Mounting the camera

You could mount the camera to the head using double-sided tape, but that method has potential problems. First, cameras often run warm, and that could weaken the tape over time. Watching a robot's eyeball fall off is a hideous thing! Second, board cameras often have a connector protruding from the back of the circuit board, which would prevent the use of tape.

Glue is another option, but not one I would suggest either. First, it prevents the easy removal of the unit, and you might want to use the camera on another project someday. Second, some types of glue can corrode circuits. A robot with cataracts wouldn't be good!

Most small board cameras I've seen have two or four small mounting holes in the corners of the circuit board. This is the most effective way to mount the camera and avoid the pitfalls mentioned with other methods.

The holes are usually very small and designed to accept #2 screws. Even if there are four holes, you need to use only two to provide strong mounting. I'll be using metal screws and nuts because nylon versions don't usually come in #2 sizes.

The mounting holes usually have a ground pad to prevent shorts, but it's often too close to the circuit components for my liking. Instead, I'll be using nylon insulating washers to prevent shorts on the circuit board.

Do not tighten the nuts against components on the circuit board too much. When complete, paint a small dab of fingernail polish on the screws to prevent the nuts from vibrating loose.

Follow these steps to mount the camera to the head perf board:

1. **Place the camera on the head perf board and mark where the mounting holes should be, as shown in Figure 16-7.**

2. **Mark the location of the connector on the perf board.**

   In Figure 16-8, I've completed marking the locations and I'm about to drill the first hole.

3. **Drill holes for the mounting hardware and the connector.**

   The board should now look like Figure 16-9.

4. **Use the machine screws to mount the camera to the head through the mounting holes just drilled. Insulate both sides of the camera board using nylon washers.**

   When you're finished, the board camera should be mounted like the one shown in Figure 16-10.

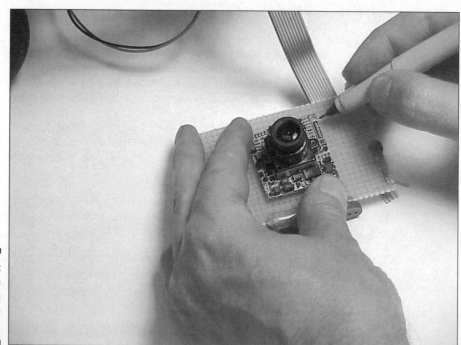

Figure 16-7:
Mark the mounting holes on the perf board.

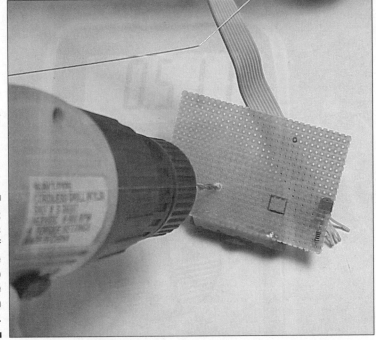

Figure 16-8:
The holes on the perf board make it easy to keep the drill in place.

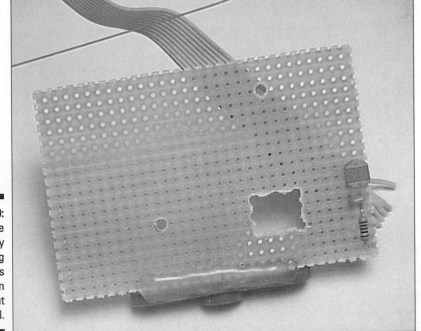

Figure 16-9:
Cutouts are created by drilling many holes and then cutting out the material.

Figure 16-10:
The board camera is mounted to the head perf board.

Providing power

Most board cameras come with a short cable harness that has a male RCA connector for the video signal and a separate circular connector to supply power. Your robot will need to supply power to the camera through the power connector.

The power connectors are usually called DC power connectors and have an outside metal shell and an inside conductor. You've probably seen them on wall-transformer power supplies used on radios and other consumer electronic equipment.

DC power connectors (seen in Figure 16-11) have no standard for voltage, size, or polarity. On top of that, some are even used to supply AC voltages. This is evident from the universal wall transformers, which have switches to select polarity and voltage, along with various adapters for different size connectors.

To make matters worse, some connectors vary only slightly in size. The main offenders are 2.1mm and 2.5mm sizes. These are often confused and result in intermittent connections. Search the data sheet for your camera; if you're lucky, it will reveal the size used. If not, you have to either guess or purchase both sizes. Luckily, they're inexpensive and the extras you buy can go in your parts bin for a future project.

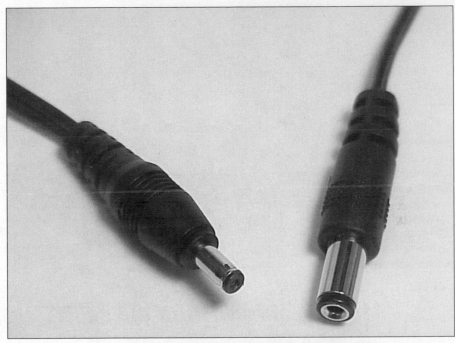

Figure 16-11:
DC power
connectors
don't benefit
from a
standard.

While looking at the datasheet, also find the polarity of the connector. Usually the center pin is + and the outside shell is ground. Applying power to the camera incorrectly will probably destroy it, so pay careful attention when wiring.

Another option is to cut the power connector off the cable and supply your own, or wire the power directly. Wiring directly might be simple, but it prevents the camera from being removed as a separate assembly, and you know I'm a stickler for modularity.

Determining connector gender

Connector gender is determined by the pins, not the housing of the connector. A connector may have a body or shell that fits inside the mate yet have a pin or pins instead of a socket. In that case, the connector would be a male. RCA connectors are a perfect example: The male version has a pin, but the female fits inside the male's outside shell. This may be difficult to envision, so refer to Figure 16-12. Connector gender is important to understand, especially when placing orders without an image.

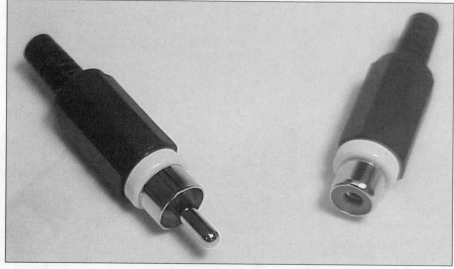

Figure 16-12:
Connector
pins
determine
gender, not
the shell or
housing.

Making and soldering the cable

After understanding and solving the connector issues, it's time to make a short cable from the source of power to the connector on the camera's cable. This can be a very short cable (2 inches) if the camera's cable is more than 8 inches long.

Power can be tapped from the bottom of ARobot's controller board near the power switch. There will be two wires: a red wire for +12 volts and a black wire for ground. Solder the red wire to the center pin of ARobot's power switch. Solder the black ground wire to pin 2 of ARobot's power connector (J9). This wiring allows you to use the power switch to turn off the power to the camera also.

After making and soldering the cable, and before connecting it to the camera, check the voltage and polarity with a voltmeter and double-check the camera's data sheet for information about its power connections. Figure 16-13 shows the power connector wired to ARobot's controller board.

When the camera is mounted and powered from the robot's batteries, check its operation by connecting the camera's video signal to a monitor. This can be accomplished with a simple RCA cable connecting the two devices. If you don't see video on the monitor, check these items:

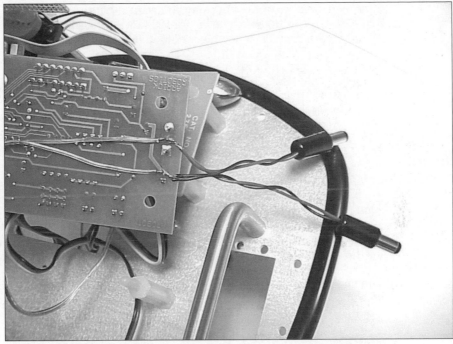

Figure 16-13:
The camera
and
transmitter
receive
power from
ARobot's
controller
board.

✔ Is the camera-mounting hardware shorting to camera circuits?

✔ Did you wire the power cable correctly?

✔ Did you remove the lens cap?

✔ Is the camera connector inserted correctly?

✔ Did you use fresh batteries?

✔ Is the power turned on?

Now it's time to make a decision about which style of system to make: tethered or wireless? When you make that decision, proceed to the next section or the one after that, whichever is appropriate.

Creating a Tethered Video System

A tethered system mainly consists of a camera, a monitor, a power source, and a bunch of cables. The cables are important and care needs to be taken with their selection and installation.

First, you need to decide how long you want the tether cable to be. A 10-foot cable is too short to be useful, and a 100-foot cable is probably too long and would create drag. I'm selecting a 25-foot cable as a compromise.

You also need to know at this point what signals your tether cable will be carrying. You can have a tether with only video, or you can add audio, data, or power.

Video-only tether

Creating a video-only tether is easy because you can get away with a single premade speaker cable. The cable should have one male RCA connector on each end and preferably have coax-style wire. Coax has a center conductor shielded by ground and provides a cleaner signal than non-coax cable. Coax is round. Non-coax has two small wires bonded together side by side — one for the signal and one for ground. See Figure 16-14.

The male RCA connectors should plug right into your camera's cable and your monitor. If not, use adapters or make your own ends.

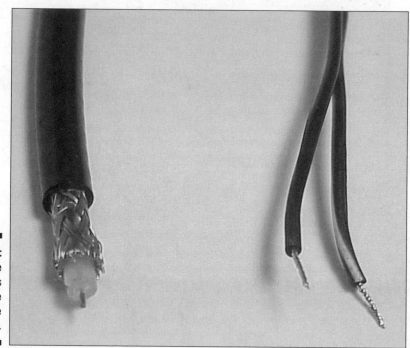

Figure 16-14:
The cable on the left is coaxial; the one on the right is not.

Tethering other signals

Tethering other signals can make complete remote control a reality. The downside is that you'll need a fat cable to carry all those signals and that will drag on your robot and reduce its maneuverability.

All the signals except power can use very small wires such as awg 24 to 28. Power, on the other hand, requires special attention.

Audio

If your camera has an audio output, or if you've added one yourself, you'll need another cable to carry that signal. You can use a small coax cable just like that used for video, or you can simply use a single strand of wire from a multiconductor cable. Wire size is not much of an issue because there isn't much current involved. Like video, audio components tend to use RCA connectors.

Data

Four signals are needed to carry data both ways (pins 2, 3, 4, and 5 on the 9-pin connector). Pin 5 is ground and can be shared with other grounds on the cable as long as they're all tied together on the robot. The signals carry very little current, so you can use small awg 26 or 28 wires. The connector on the robot side is a 9-pin D-sub male, and the connector on the computer side is a 9-pin D-sub female. Use the schematic in ARobot's user guide for additional information.

Power

Because ARobot can draw as much as 1 amp under a heavy load with servo motors and other items going, the wires that carry power should be larger. I suggest two awg 22 wires to carry +12 volts and ground to the robot. If necessary, you could use two or more smaller wires. The other side of the cable could be connected to a bench power supply to save batteries and provide unlimited robot life.

Individual cables

You can use individual cables to carry the various signals: one for video, one for audio, one for data, and one for power. That will be a big bunch and you'll have to tape them together to prevent a tangled mess. A bunch this large is likely to drag on the robot quite a bit, so make sure the cables are as small as possible.

Multiconductor cables

Instead of using a large bunch of cables, you could use a single multiconductor cable. This can be a round cable with an external sheathing or a small flat cable.

For the large round cable, you could use a 9-pin serial cable with an adapter cable on each end to break out the individual signals, or you could purchase a length of raw cable and make your own. I suggest 6- or 8-conductor cable with awg 24 stranded wires.

Flat cable is lightweight but isn't designed to carry much current. Usually the wires are awg 28, which is fine for all signals except power, which should have two wires for +12 volts and two for ground. A 10-conductor flat cable is a standard size and can transmit everything you need. Like other cable, it's sold by the foot or in spools of 100 or 1000 feet.

An interesting option is to use phone line cable. I'm not referring to stiff, solid wire, but the small flexible wire intended for use between the wall plate and the phone. It comes in 4- and 6-conductor style and is capable of handling all the signals except power, unless you can dedicate two or three wires each to power and ground.

Attaching cables

The final issue is often overlooked and underappreciated: mounting the cable to the robot. The cable should be attached to the robot so that it doesn't interfere with the steering linkage or whiskers. Mounting should also provide a strain relief. Ideally, the robot should be able to hang from the cable without the cable breaking or coming loose.

One option for mounting the cable is to use a spring-loaded bar that holds the cable 6 to 12 inches above the robot and provides some give-and-take for the cable. See Figure 16-15. This type of bar is used to hold cords for ironing boards to prevent an accident.

Another option is to simply attach the cable to the base of the robot using a plastic wire tie or a cable clamp, as shown in Figure 16-16. When complete, it's important that a pull on the cable doesn't tug on the camera or other electronics.

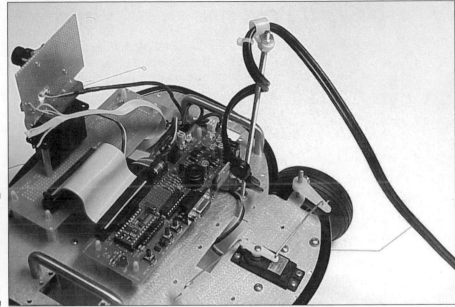

Figure 16-15:
Mounting
the cable
with a
threaded
rod clamp.

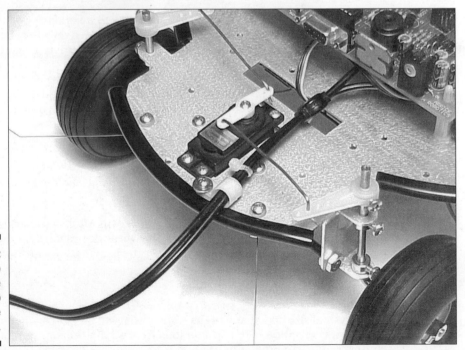

Figure 16-16:
You can also
use a cable
clamp to
secure the
cable.

Using the tethered video system

Before powering up, make sure any metal parts from the connectors are not touching the robot's controller board. Secure them or wrap them in electrical tape if needed. After the cable is mounted and all connections are made, power up the monitor and the robot to make sure the video system is working.

It might be necessary to adjust the camera's lens for a good view. Usually you have to loosen a small screw first. Then you can rotate the lens to improve focus. Some lenses allow this and some don't.

The robot should operate in an open area with little impact from the cable, but going around objects will be difficult because the cable will try to hang. Sometimes the robot may move so that the cable actually triggers one of the whisker wires. If the cable is very long, you might discover that the robot struggles with the extra drag. There's no solution to these problems except to go wireless. The next section shows you how.

Creating a Wireless Video System

Without wires, your robot will feel free, and so will its operator. Without the restraint of an electronic leash, you'll be able to maneuver the robot around corners and through furniture legs, and the robot can travel much longer distances.

In a nutshell, a wireless video system includes a camera and a monitor just like a tethered system, but it also has a transmitter module and a receiver module. These two items effectively replace the cable that would ordinarily connect the camera to the monitor, thereby eliminating the robot's leash.

Installing the transmitter

The transmitter will have to mount somewhere on ARobot's chassis so that cables from the camera can reach it, power wires can reach it, and the antenna can stand. It's also important that the transmitter's location doesn't limit access to ARobot's controller board.

The physical part

The physical size and shape of the transmitter will influence where it can be mounted. One option is to use standoffs to raise the transmitter above the controller board, and another option is to mount it underneath the controller. Figure 16-17 shows how I simply slid the transmitter under the controller board after connecting the cables.

Figure 16-17:
Small
transmitters
can be
mounted
under the
controller
board like
this.

Supplying power

Choose a transmitter that runs on 12 volts DC. That way, ARobot's battery pack can supply power. Make a cable just like you did for the camera but with a connector that matches the transmitter.

Connecting the receiver and monitor

You need a cable to connect the receiver to the monitor, as seen in Figure 16-18. This will probably be a short cable with male RCA connectors. Receiver modules are usually powered with an AC wall adapter.

Using the wireless video system

Operation of the wireless transmitter and receiver is easy because it simply emulates a cable. If you've tested your video system with a cable and then replaced the cable with a working wireless system, everything should work fine.

For testing, load the wander.bs2 program into ARobot and let the robot move around your house. (For more on the wander.bs2 program, see Chapter 10.)

Move the video receiver and monitor to various locations to gain an understanding of the performance and possible locations of interference. Also, experiment with various positions of the antenna (if there is one) and orientations of the receiver.

Figure 16-18:
The components on the receiving end are easy to connect together.

Troubleshooting

Wireless systems are not without pitfalls. If your system has fallen into such a pit, check out this list for some troubleshooting hints:

- **Power:** Every device in the system — the monitor, receiver, and transmitter — needs power. The camera was powered up and tested earlier. Make sure that each device is getting the correct voltage and has the proper polarity. Don't forget to test each power source with a multimeter while power is on and the load is present. Simply use your multimeter on the DC volts scale (10 or 20 volts will do), connect the black lead to ground and the red lead to the point that you want to measure.

- **Channels:** There may be channels to select depending on the type of equipment. The transmitter and receiver must be on the same channel;

if the units use TV frequencies, select a channel that doesn't interfere with local stations.

✔ **Connectors and cables:** If you made the connectors and cables yourself, they are the most likely cause of any problems. Not to question your soldering skills, it's just a fact of life. Make sure that each connector is soldered correctly and that each wire goes were it's supposed to.

✔ **Interference:** A common problem with wireless links is interference from nearby electrical devices, or reflections and blockages caused by metal walls. Computers and fluorescent lights are the worst offenders, but devices with large motors such as garbage disposals and refrigerators can also wreak havoc on wireless transmissions. Move your receiver to a different location if you suspect interference and make sure your antenna (if it exists) is connected correctly and pointed up if possible.

Wireless Data Links

Because this chapter deals with wireless video, I thought it relevant to talk briefly about wireless data links. Wireless technology is changing fast and new products are introduced almost monthly. Prices are also changing too, making it more affordable all the time to communicate without wires.

You can use ARobot's serial communications port to control the robot and to receive data from its sensors. A simple program can be written to wait for characters at the serial port and respond by moving motors, reading the whiskers or light level, or even starting other programs with specific functions, such as making an alarm sound. The program could display the status of the robot's sensors and the robot's position on your computer screen in real time — now that sounds like fun!

Several companies are producing wireless data link products that can transmit and receive data from serial ports. Many of these units operate on 12 volts just like your camera transmitter and are perfect candidates to remotely control your robot.

If you're interested in making your robot communicate without wires, pick up a current electronics magazine and look at the ads. Chances are, you'll find a product that will set your robot free forever. Figure 16-19 shows a typical wireless data link system.

Figure 16-19:
Wireless
data links
allow
complete
remote
control.

Camera Positioning

Because the camera is located on the head perf board, the robot can control the viewing angle. Just like a human head, the robot can move its head for a better view instead of wasting power moving the entire body. Moving the head is also much faster.

The RC servo motor used on the head is designed to move approximately 60 degrees on each side of the center position. In reality, RC servo motors can usually move a full 90 degrees on each side of the center position. Either of these figures will give us plenty of movement. Exactly how far your robot's head will move can be determined by experimentation.

In this section, you find out about program code that you can use to move the head in a logical fashion. ARobot's coprocessor handles the underlying complexity of driving the motor. So, to move the motor to the desired location, you just send a simple command to the coprocessor. The details of commanding servo motors are described in ARobot's user guide.

Facing straight

The most basic function is to make the head point directly ahead. RC servo motors do not move to a position unless commanded to, so if you want the head to look forward, you have to specifically request that from the coprocessor. The following program snippet can be incorporated into a larger program. Place it at the beginning of most programs that use devices attached to the head.

```
serout 8,396,["!1R280"]   'Move head center
pause 300                 'Wait for completion
```

The first line sends a command to the coprocessor telling it to move RC servo motor #2 to position 80. The command is sent using the serout command, which sends the data in a serial fashion. The 8 after serout specifies which port pin is used to communicate with the coprocessor, and 396 specifies the data rate. You can find complete information about the serout command in any Basic Stamp 2 programming documentation text. The second line simply waits a short time to allow the motor to move to the requested position.

The string of text to send to the coprocessor is shown in quotation marks. The !1 specifies coprocessor 1 (there can be others), and R2 specifies RC servo motor 2, followed by a two-digit position code, where 80 is the center, 01 is far left, and FF is far right. Sending 00 as the position relaxes the servo motor, saving power and allowing it to spin freely. Otherwise, the motor will attempt to maintain its position resisting any forces. These position numbers are in hexadecimal, which is described in most computer programming texts as well as ARobot's user's guide.

Three-step panning

It's helpful to have the robot pan the camera in a logical fashion to acquire a larger field of view. You can accomplish this by panning to the left and right of center by an amount that makes the three frames connect to produce one larger view for the operator.

Exactly how wide the camera's view is depends on your camera's lens. Experimentation will show you what positions will produce frames that connect.

The scanh subroutine in Listing 16-1 can be called to pan the camera left, center, and then right. A short delay is inserted at each position, and the head is positioned to the center and then relaxed when the sequence is finished. Comments in the subroutine describe the operation. The beginning of the listing is a short program that sets constants and calls the subroutine.

Listing 16-1: Scanning

```
'Scanh.bs2 - Roger Arrick - 5/23/03
'Short test program for the scanh subroutine
net     con 8               'Coprocessor network pin
baud    con 396             'Coprocessor baud rate

Start:  Gosub scanh         'Scan the head
        pause 3000          'Wait 3 seconds
        goto start          'Loop forever


'-------------------------------------------------
'scanh
'Scans the head left, center, right with delay.
'Ends with head centered and relaxed

scanh:
        serout net,baud,["!1R230"]   'Move head left
        pause 2000                    'Pause 2 seconds

        serout net,baud,["!1R280"]   'Move head center
        pause 2000                    'Pause 2 seconds

        serout net,baud,["!1R2D0"]   'Move head right
        pause 2000                    'Pause 2 seconds

        serout net,baud,["!1R280"]   'Move head center
        pause 200                     'Pause .2 seconds

        serout net,baud,["!1R200"]   'Head motor off

        return                        'Done
```

Slow scanning

The program in Listing 16-2 shows how to perform a slow scanning operation so that the operator can get a larger view. RC servo motors are designed to go to each position as fast as possible, so this code has to move to each position sequentially with small delays between each step. The result is a slow motion from left to right and then back to the center to rest.

Listing 16-2: Slow Scanning

```
'sscanh - Roger Arrick - 5/23/03
'Short test program for the sscanh subroutine
```

```
net      con 8               'Coprocessor network pin
baud     con 396             'Coprocessor baud rate
hp       var byte            'Head position

Start:   Gosub sscanh        'Scan the head slowly
         pause 3000          'Wait 3 seconds
         goto start          'Loop forever


'-------------------------------------------------
'sscanh
'Scans the head left, center, right in a continuous
'slow motion. Ends with head centered and relaxed.

sscanh:
         For hp = 1 to 255    'Loop through positions
           serout net,baud,["!1R2",hex2 hp]  'Move head
           pause 50           'Short delay
         next

         serout net,baud,["!1R280"]  'Move head center
         pause 200                    'Pause .2 seconds

         serout net,baud,["!1R200"]  'Head motor off

         return                       'Done
```

Adding a tilt axis

After seeing the view from your new motorized camera, you might be thinking that adding another axis would further increase the usefulness of the system — and you'd be right! Dual-axis systems like that are referred to as *pan and tilt* and are often found on security cameras and in professional studios.

ARobot's current head motor provides the pan axis, or left-to-right horizontal motion. To add a tilt axis (up and down), you need another servo motor. ARobot's controller board has extra servo-motor control ports, so the most difficult part will be the mechanics of attaching the new motor so that it provides the desired movement.

Figure 16-20 shows my attempt at adding a simple tilt axis to the existing pan axis.

Figure 16-20:
A system with pan and tilt improves the view of your new video system.

Applications Using the Video System

Now that your video system is complete and working, it's time to consider how you could use the camera and video information in a real-world application. Because the robot's computer doesn't have access to the video information, it can't be used to make decisions, but the robot does have sensors that could be used to alert the operator to check the video for activity. In this section, I toss out some ideas about how to make use of the new system.

Navigation

Of course, the video information can be used to provide feedback to the operator for navigational purposes. This assumes that the robot can get, and respond to, commands from the operator, as in the case of a tethered or wireless data link. The control program could also allow the operator to change the view using the head motor.

Inspection

I've seen several shows on TV that used small robots to inspect remote locations where a human couldn't fit. A good example was the exploration of a long vent tube in the great pyramid. Their robot was tethered and had a small, high-quality video camera on the front. Too bad they didn't find the great hall of records!

The pyramid exploration application is similar to industrial applications such as pipe inspection. Often, pipes that carry gas or some other material need to be checked for cracks and damage, and a remote-controlled robot is the perfect solution — especially when a human would be risking his or her health.

Another possibility is to use the robot in a place where a human shouldn't go. A good example of this is a nuclear power plant or a chemical plant where the air or radiation might prove dangerous to a worker.

Exploration

More and more, we're seeing robots perform exploration in places that are too dangerous for humans. A good example of this is the exploration of volcanoes where temperatures are very high. Tether cables are the norm in earthly applications such as this because robots tend to get stuck and researchers don't like the idea of leaving a multimillion-dollar machine in a hole. Tether cables also allow plenty of high-speed data collection.

In recent years, wireless robots have been used in space exploration, specifically on Mars. The driving force for this is cost, but human safety is a factor as well. Because data transfer time over long distances of space can be minutes or hours, the robot must perform some actions without direct human intervention.

Security

As you can imagine, security is one of the fastest growing applications for automated machines. I believe in the future it will become commonplace for remote-controlled robots to roam the halls of our schools, businesses, and homes, checking for security issues. Intelligent sensors will play an important role in helping humans detect certain situations.

Your ARobot can already perform some of these security functions. For example, using a motion detector, it can trigger video scanning when a human walks. The speaker or speech synthesizer can be used to communicate or alert those nearby. ARobot's camera could be used to scan the underside of vehicles or stand guard on a porch.

A Robot's Vision

Although your new video system doesn't perform a programmed vision function, the amount of fun and usefulness in simply transmitting video back to the operator should not be underestimated. Some great adventures await your newly enhanced mechanical friend. The possibilities are truly unlimited — that's what I like about robotics.

Chapter 17

Controlling Your Robot from Afar

. . . .

. . . .

Remote control is a concept any couch potato can cozy up to. With remote control you can flip on the TV from across the room or turn the lights down low while sitting on the sofa with your sweetie. You can be the pilot of your RC boat or plane. With remote control on your robot, you can do things like sending your robot into dangerous terrain while controlling its every move from a safe distance.

In this chapter, I show you how to make your ARobot remotely controllable using a standard, off-the-shelf system designed for hobbyists. With this remote-control system, ARobot's controller can even retain a certain amount of autonomous control to provide the best of both worlds.

All about Remote Control

Out of the box, ARobot is a pretty independent type of guy. But even though ARobot is set up to act autonomously through the programs you download to its controller, you can add remote-control functionality to it to give you a more hands-on robotic experience.

Technically, *remote control* simply means that an operator can control a device from a distance. This control can be through a tether cable, *infrared* (IR) connection, *radio-frequency* (RF) transmission, or some other means. In this section, I give you an overview of remote-control technology and explain which is the best RC system to use in robotics applications.

Figure 17-1 shows ARobot controlled by a remote-control device.

Figure 17-1:
ARobot
being
remotely
controlled.

Optical remote control

Optical remote-control systems are just another term for infrared technology. You probably have several infrared devices cluttering your coffee table. Optical systems such as IR systems are best used in situations where the operator is in view of the machine at all times because infrared signals don't transmit easily around obstacles or corners. A good example of this type of device is your TV remote control.

Because you may want to send your robot out of sight or even out of the room at times, infrared probably isn't your best option.

Tethered remote control

Like a hi-tech leash for your robot, a *tether* is a cable that transmits a signal that allows you to control your robot's actions and sometimes the power. Tethers are by far the easiest RC systems to build and often the most reliable, and nothing beats a tether for signal strength. Tethers also aren't plagued by the electronic noise that can interfere with signals that have to travel through the air.

But the obvious limitations of a long cable running between you and your robot prevent it from being the system of choice except in certain situations, such as volcano exploration. Having a tether in that application is an advantage because it allows the operator to yank the robot out in a hurry if a volcano suddenly decides it's a good time to erupt.

Radio-frequency remote control

Radio frequency is by far the preferred means to remotely control a robot or other machine because you're not limited to a line-of-site as you are with an IR or other optical system, and you're not encumbered by the dangling cables of a tethered system.

Radio-frequency transmissions, such as those used by a garage-door opener and cell phone, work by transmitting an electromagnetic wave through the air just like your local radio station. RF signals can go through buildings and can even travel around the world if they're powerful enough.

Nothing's perfect, however, and RF systems are not without their problems and limitations. The main issues are interference and range.

Interference to an RF signal can come from any device that emits electrical noise, such as a fluorescent light, a refrigerator, or another transmitter. Interference can cause an intermittent signal or a lost signal.

The *range* of an RF system — that is, the size of the area it can transmit across reliably — is not unlimited. Range can change based on obstacles between the transmitter and the receiver, atmospheric conditions, and electrical interference. Sometimes, having a transmitter too close to a receiver can also cause a problem due to overpowering (distortion due to a very large signal) or local reflections (waves bouncing around creating conflicting signals).

In spite of these issues, RF is the system I prefer to use to control ARobot because it provides the ultimate freedom of movement.

The Lowdown on RC Systems

To be a smart consumer, you should know some things about how RC systems work before you buy one. In this section, you find out about the options available to you and become a savvy RC system shopper.

Model RC systems

If you've been in a hobby store lately, you've seen a wide variety of model airplanes, helicopters, cars, and boats, all designed to be controlled with an RC system. RC systems have become a big business and many different units are available.

Usually, an RC system for a model is sold separately from the vehicle. Several types of RC systems are available, each made for specific types of vehicles. Systems for cars often have a small steering wheel, and systems sold for airplanes have a joystick, for example.

A typical RC system includes a battery-powered transmitter box that contains the transmitting circuitry, joysticks for controlling the motors, and an antenna. The system also has a small receiver box with connectors for the motors along with a wire that acts as an antenna. The motors (two to six in a typical system) are small, box-shaped devices with a shaft that rotates and a short cable that connects to the receiver. The system may also include special batteries and chargers.

The number of motor channels in a remote-control system varies depending on the vehicle it controls. Each motor channel consists of a control such as a joystick or switch, and a servo motor. A car often needs only two motor channels: one for steering and one for throttle. An airplane may have three, four, or even more motor channels used to control the various devices needed to make it fly. Figure 17-2 shows a typical RC system used with airplanes. Basically, the more money you spend, the more motor channels you get and the more reliable your signal will be due to fancier circuitry.

Figure 17-2:
A typical four-channel RC system comes complete with servo motors.

Different types of transmission schemes are used in modern RC systems, including PPM, PCM, and IPD. Don't let all these initials confuse or concern you too much. The systems are designed with matching transmitters and receivers, and the underlying technology is something you never have to tinker with. As long as you don't mix transmitters and receivers from different systems, you shouldn't have a problem.

RC servo motors: The power behind RC

Like many mechanical critters, an RC system requires a motor to do things. One thing that the different RC systems have in common is a fairly standard servo motor and interface (connector and signal definition). The beauty of RC servo motors is that the entire motor-control system is inside a small plastic block, so the motor is a self-contained module. This module has a short cable to connect it to the device that will control it and an output shaft used to move your device. Inside the module are gears, a motor, and a control circuit.

Figure 17-3 shows a typical RC servo motor complete with the hardware used to mount it and horns used to connect the output shaft to the model (or in our case, the robot).

Figure 17-3:
RC servo
motors
come in
various
sizes.

Many sizes of servo motors are available, but they're all controlled the same way: using a 3-pin connector that plugs into the receiver module. Figure 17-4 shows this connector up close. This servo-motor interface standard has accomplished a lot for robot builders because it's made servo motors easy to control and easy to purchase. Also, servo motors have become very inexpensive for all that you get.

Figure 17-4:
The beauty
of an RC
servo motor
lies in its
interface
connector.

The RC servo-motor interface

Here's info that all of you who enjoy knowing how motors work will like. The RC servo-motor interface connector has three signals: power (usually 4.8 to 6 volts), ground, and the control signal. What's really interesting about this setup is the control signal. It's a digital pulse whose width determines the position of the motor. When not receiving pulses from a receiver or controller such as ARobot's, the motor just sits there, but when receiving pulses, the motor moves to the desired position and stays there, resisting external forces.

The overall range of travel for most RC servo motors is about 120 degrees (plus and minus 60 degrees from center).

The reason this interface is so useful for robot builders is that it's easy to control with a microprocessor. The pulses are not very fast, so even a slow processor has enough time to send out pulses to control many motors, sometimes as many as eight or more. ARobot's controller board can talk to four RC servo motors; one of which is used for steering and the others for accessories such as a head or gripper.

RC system frequencies

Radio-frequency RC systems operate within frequency bands created specifically for them. These bands are broken down into channels.

It's important to know that flying RC devices have different frequency bands than RC vehicles that move around on the ground due to safety concerns. Who wants to be flying an RC airplane only to have it crash because Joe hobbyist a mile away is playing with his RC monster truck?

For your robot, stick to a system made for use on the ground. Also, if your robot will be operating in the vicinity of other devices using RC systems, make sure they're using different frequencies. Otherwise, interference will ruin your whole day. The frequency can usually be set by changing a crystal or setting a switch, so see the owner's manual for more information.

Channels

The term *channel* refers not only to the units that frequency bands are broken down into but also to the number of servo motors that can be controlled by the RC system. This may sound confusing (probably because it is).

Most RC systems have at least two channels; some have as many as eight or ten. For this project, you'll be using one servo for steering, one for camera panning, and one to control the drive motor, for a total of three. With this setup, you need a three-channel RC system at a minimum. Your robot can read additional channels, thus allowing you to control other things such as speech or a headlight.

Purchasing an RC System

It's time to find the perfect RC system and fork over the bucks to make it yours. Most hobby stores and online vendors of model airplanes and cars offer RC systems, usually listed under the category of radios. If you live in an area that has plenty of strip malls, chances are you have access to a hobby store that carries what you want, so plan a shopping trip. But before you go, here's some advice from a veteran RC system shopper.

What's the cost?

RC systems will cost you anywhere from $60 for a low-end, dual-channel unit to more than $1000 for a multichannel unit with a digital display and other fancy features. As with any purchase, you can go overboard. I suggest sticking with a simple four-channel unit made for surface vehicle use. You can get one for around $150 complete with transmitter, receiver, battery packs, and servo motors. That's just fine for this project and most robot applications.

Buying online

I prefer to see what I'm buying when possible, and nothing beats a retail store with an educated salesperson. If that's not an option for you, consider purchasing your system online. You'll probably get the best price that way, but don't expect to get your questions answered by a live person (at least not promptly). Look for an online retailer that provides product specifications and FAQs (Frequently Asked Questions) on its Web site.

Here's a list of some online dealers of RC systems. These vendors offer a wide selection of systems from various manufacturers, along with secure online order forms:

Company: Roy's Hobby Shop

Web site: www.royshobby.com

Offerings: Extensive line of RC systems from two to ten channels and a variety of features. Reasonable prices and a knowledgeable staff. This is an online store that will answer your questions! They also have a large retail store in Hurst, Texas.

RC System: VG400 Vanguard 4-102Z FM 75MHz by AIRTRONICS

Price: $134.99

Company: Hobby Warehouse

Web site: www.hobby-warehouse.com

Offerings: Complete line of model aircraft, boats, cars, and so on. Also carries science kits and other related products. They have a retail outlet in Kentucky.

RC System: Hitec Ranger III

Price: $79.99

Company: Tower Hobbies

Web site: www.towerhobbies.com

Offerings: Everything from model rockets to model airplanes. Also model trains and tons of accessories for the hobbyist.

RC System: Futaba 4VWD

Price: $109.99

Installing an RC System

Adding an RC system to your ARobot is a pretty simple process. You attach the RC receiver to ARobot's body, and then connect wires from the receiver to the servo motors and also to the expansion board. The receiver can rob power from the +5 volt rail, so you don't even need to add extra batteries.

After you've installed the RC system, I'll show you how to write a short program to read one of the RC channels and make it control the robot's drive motor. Along with a wireless video system, you'll also have a powerful remotely-controlled machine that you can use to collect images from a remote location.

Wiring

The first thing to tackle is the wiring. You'll use three of the RC channels to control the robot: one for steering, one to pan a video camera, and one to control the drive motor.

You can use the fourth channel for a tilt axis, if you happen to have one on your camera.

Steering and camera cables

Wiring for the steering and camera servo motors is a matter of switching out connections. You just move the motor connectors from ARobot's controller over to the RC receiver, as shown in Figure 17-5. How you connect these connectors will depend on your RC receiver, so see the documentation which will describe which connector matches which channel and the orientation of the connectors.

With these connections switched, the motors are now no longer under program control, but instead are under the control of the RC system.

Connecting the drive motor cable

The next step, connecting the drive motor cable, involves creativity with cables. You have to rig a cable that plugs into the receiver on one side and connects to the expansion perf board on the other. A good place to start is with a short servo-motor extension cable that has a male connector on one end and a female connector on the other.

Extension cables can be purchased from most hobby shops that sell RC systems.

To make the cable, cut off the male connector to expose the wires on that end. You'll connect one of these exposed wires to the expansion board, and connect the female connector on the other end to the RC receiver.

Figure 17-5:
Connecting
the steering
and camera
motors to
the RC
receiver is
simple.

You have to deal with only one wire on this connector — the servo signal itself, which is usually white and sometimes yellow. The other two wires (red and black) can be cut off and ignored.

Now that you have the cable ready, follow these steps to connect it:

1. **Strip the servo control wire to expose $\frac{1}{16}$ inch.**

2. **Tin the exposed wire with a dab of solder.**

3. **Solder the tinned wire to pin 11 of the expansion connector (on the bottom of the perf board), as shown in Figure 17-6.**

4. **Insert the connector into the receiver module, as shown in Figure 17-7.**

Power cable

Now you'll get involved in a bit of power thievery. The receiver module is designed to get power from a 4.8-volt rechargeable battery pack supplied with your RC system package, but that would add unnecessary weight to your robot. Instead, you can steal power from ARobot's expansion board.

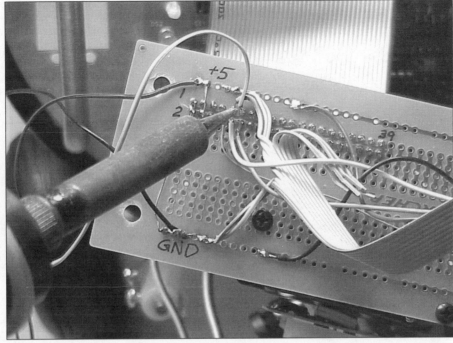

Figure 17-6:
One
channel
of the
receiver
connects
to the
expansion
board.

Most RC receivers can operate fine on the 5 volts that your robot supplies, but it wouldn't hurt to check the RC system's data sheet before you buy, just to make sure.

The receiver has a special connector (often conveniently labeled B for battery) that transmits power. The connector is usually the same type as the servo-motor connectors used in the preceding section, but it has only two wires. In addition to the receiver connector, just as with the drive motor connection, you'll need a cable and connector to go from the receiver's battery connector over to the expansion board to get power. I suggest that you start with a servo-motor extender cable and chop off the male end to create the cable you need.

The middle wire of the cable is likely to be red and should be wired to the +5 volt rail on the bottom of the expansion perf board. The other wire, which is ground and probably black, should be wired to the ground rail. If you're using a servo-motor cable, there will be a third wire (probably white or yellow) that you can simply cut off.

Figure 17-7:
The RC receiver with the servo motor and power cables attached.

Follow these steps to connect the wires:

1. **Strip the end of both wires to expose about ¹/₁₆ inch.**

2. **Tin the two wires with a dab of solder.**

3. **Solder the red wire to the +5 volt rail on the bottom of the expansion board and the black wire to the ground rail, as shown in Figure 17-8.**

4. **Insert the connector into the RC receiver, as shown in Figure 17-9.**

I suggest you do one last check of the wiring and soldering before turning on the juice. Accidentally swapping power and ground might destroy your receiver and result in a very sad robot. (Have you priced robot antidepressants lately?)

Mounting the RC receiver

For your robot to receive RC commands, you have to mount a receiver on it. RC receivers are small and lightweight because they're made for model airplanes, cars, and other vehicles that have light payloads.

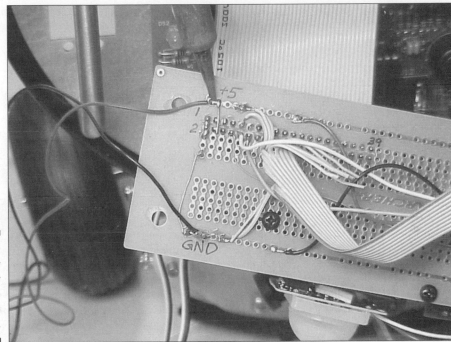

Figure 17-8:
The receiver gets power from the expansion board.

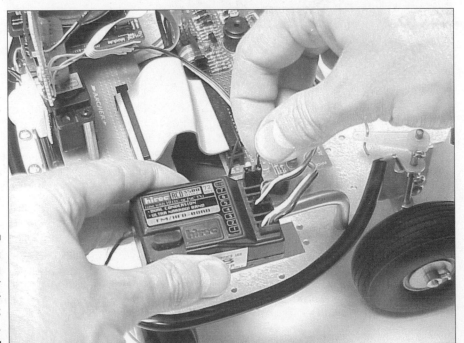

Figure 17-9:
Inserting the power connector into the RC receiver.

I suggest mounting the module using a small piece of Velcro with adhesive. You can mount the unit anywhere on the robot where there's an empty space. I decided to attach the receiver on the underside of the body plate near ARobot's battery pack, as shown in Figure 17-10. There's plenty of room down there and the wires can easily go through one of the openings in the base plate.

Figure 17-10:
Using Velcro allows the receiver to be removed if necessary.

Antenna

Just as those little RC model cars you've seen kids playing with in the park sport an antenna, your robot has to have an antenna to pick up radio-frequency signals from the RC unit. The antenna for most RC receivers is simply a wire. In model airplanes, the wire is strung between two surfaces; in model cars, it's often stuck in a clear tube to keep it straight and mounted pointing up (just like on a real car) for the best reception. Plastic tubes made for this purpose may come with your RC system or can be purchased at a hobby store that sells RC system accessories.

I've found that the best option is to use an antenna tube and hot glue it pointing up (Figure 17-11 shows me doing this). If you'll be using your system mostly within close range, it may be okay to just wind the antenna wire around the handles or another fixture on the robot, but keep in mind that this is an inexact science and you'll have to experiment to see what position provides the best reception.

Figure 17-11:
The antenna tube is glued to ARobot's base plate.

Be careful with a hot glue gun. Otherwise, you might melt the antenna wire.

Testing

If you're confident in your wiring skills (especially that you wired the power connections correctly) and you've mounted the receiver, you're ready for the acid test. Turn ARobot's power switch on and see what happens.

What *should* happen is not much, except, possibly, that some servo motors will vibrate. To get some action, power up your RC transmitter and start moving the controls. The servo motors should begin to move.

Software

You don't need much software magic to make this project work. The software just reads one of the servo channels and controls the drive motor according to those readings.

Matching things up

Connectors on the receiver match up with controls (usually joysticks) on the transmitter. When you start working with the RC device, you may find you have to move the cables around to different connectors so that the appropriate control is assigned to the corresponding servo motor.

A typical four-channel (four servo-motor) RC system has two joystick controls on the transmitter. The joysticks can be moved left and right and up and down. The steering motor is controlled with a left-and-right control if the transmitter has joysticks or with a steering wheel. The other left-and-right control should be used to control the panning (left-and-right movement) of the camera. The up-and-down control associated with the steering control should be used to control the drive motor (but that won't operate until you download the software).

The result of all this is that one of the joysticks controls the robot's motion. Left-and-right controls steering, and up-and-down controls forward and reverse. You'll find this intuitive. The other joystick is used to control the camera. Left-and-right is for panning, and up-and-down is for tilting.

The control on the transmitter should be an up-down joystick motion so that operation is intuitive. Up makes the robot move forward, down makes the robot move in reverse. Listing 17-1 shows the program that performs this simple task. You can download this program from the Web at www.robotics.com/rbfd.

Listing 17-1: Remotely Controlling the Drive Motor

```
'{$STAMP BS2}
'-------------------------------------------------
'rcontrol.bs2 - Roger Arrick - 5/23/03
'
'This program reads an input (P6) from an RC
'receiver and uses it to control the drive motor.
'The steering motor and other motors such as camera
'control can be driven directly by the RC receiver.

rcin      var word          'Pulse width from RC
net       con 8             'Coprocessor network pin
baud      con 396           'Coprocessor baud rate

here:     if in14=1 then here   'Wait for button

start:    pulsin 6,1,rcin   'Read the RC
          if rcin < 500 then stp  'Stop if out of range
```

```
          if rcin > 1000 then stp 'Stop if out of range
          rcin=(rcin-500)/50 'Scale down value to 0-10
          if rcin>7 then gof 'Forward
          if rcin<3 then gor 'Reverse

stp:      serout net,baud,["!1M1100000"]  'Stop motor
          goto start            'Keep doing this

          'Forward
gof:      serout net,baud,["!1M116FFFF"]  'Drive motor forward
          goto start

          'Reverse
gor:      serout net,baud,["!1M106FFFF"]  'Drive motor reverse
          goto start
```

After you download this code to the robot, the program begins running. When you press switch 1 (SW1), the program looks at the receiver's channel signal and controls the drive motor (forward and reverse). If the RC transmitter is turned off, there's no telling what will happen (the robot will probably behave a little . . . well . . . erratically). This is caused by the RC receiver not getting any signals from the transmitter.

At this point you'll be able to tell which control operates which motor by operating the remote-control device and observing the results. If you're using a transmitter with two joystick controls, it's best if you have the joystick left-right motion control the steering motor and the up-down motion control the drive motor. With this setup, the joystick, when moved to the right, should control the camera panning motor as well as the tilt (up-and-down motion) motor, if you have one. To change which control operates which motor, simply switch around the connectors on the receiver module.

It may be necessary to reverse the control response if a remote-control operation is the opposite of what you expect. Many transmitters have switches on the back or bottom that allow you to reverse a channel's response; see the operator's guide that came with your RC system to find out how to reverse the channels.

Some transmitters allow you to make adjustments that set the amount of travel (distance moved) for each channel's motor. The adjustment is usually accomplished with a trimmer that can be turned with a small screwdriver. *Trimmers* are small controls — usually levers — for making small adjustments to the joystick's response. To make ARobot work best, set the adjustments for full travel of each channel's motor. See the user guide for your transmitter for specific details about travel.

Troubleshooting

Not much can go wrong if you have a working RC system that's already been tested before installation. However, here are a few ideas to explore if you run into trouble:

- **Batteries:** The RC transmitter must have fully-charged batteries. Without proper voltage, the signal will not be as strong, and crazy robot action such as running over your cat can occur. The receiver could be stealing power from ARobot's power source, so make sure the robot's batteries are charged too.

- **Antenna:** The receiver's antenna is usually a long wire. It's possible for this wire to be placed in such a way that ARobot's controller interferes with its operation. The result will be erratic behavior of the motors. Change the positioning of the antenna and test to see what position works best. Unfortunately, there's no secret formula to rescue you from the black magic of radio interference.

- **Program typo:** If you typed the program into the editor yourself, double-check for typographical errors. Every character counts. The result of a typo in this case would be incorrect drive motor operation rather than problems with the steering and camera motors.

- **Wiring:** It's important that each wire go to the correct place. Compare your connections with the figures in this chapter and recount those connector pins. Also, check for shorts to nearby pins.

- **Interference:** Another nearby signal might be interfering with your transmitter's signal, although this is unlikely. I'm unaware of any product that could cause this problem, but check cordless phones, baby monitors, and the ham radio operator next door.

Half and Half: RC and Autonomous Behavior

Here's a project to put your RC-enabled robot to work. ARobot's steering motor and camera motor are controlled directly by the RC receiver, but the drive motor is controlled by ARobot's processor and program based on the signal it gets from one of the RC receiver channels. This configuration prevents the program from taking control of the steering and camera motors, which might become a limitation if you want to have ARobot operate autonomously.

Using remote control

Many applications can benefit from having remote-operator control. Here are some fun ideas to get your neurons jumping.

Spying: Why send a perfectly good human into harm's way when a robot could do just as good a job? Snooping into nooks and crannies you don't want to enter is a good example of how robots can protect their humans from dangerous situations. A robot could move around the crawlspace beneath your house under remote-operator control, for example, collecting video and audio information that's fed into a recorder. Later on, in the comfort of a warm, dry room you can review the data to calculate exactly how much dry rot you have got down there.

Remote hostage negotiator: You've probably seen them on the nightly news — a robot goes up to a house where a distraught (and toothless) guy with a weapon is threatening his goldfish with bodily harm. A worried neighbor calls 911 and the swat team arrives. Do they send in their best and brightest to deal with toothless Sam? No, they send in a remote-controlled robot with a two-way audio system. (As a variation on this, you could send that same robot into the next room to negotiate with your kids about their bedtime through the two-way audio system.)

Dangerous materials disposal: If your dog brings home dearly departed squirrels, what would be a good way to get rid of the carcass? Remote-controlled robots are the perfect machines to perform the task of investigating and disposing of odious objects. Such robots have grippers and long arms that allow them to grab an object and place it inside of a container for safe disposal.

Challenging environment exploration: Many dangerous environments should be left to robots, including nuclear waste sites, hostile planets, and your kids' bedrooms. Remote-control robots are great for surveying such terrain because they give the operator complete freedom to do a bit of exploration without worrying about endangering themselves or other family members.

Inspecting: Have you ever had to crawl around a sewer or your own attic? Wouldn't it be cool if a robot could do that? How about inspecting a long length of pipe or the inside of a narrow tunnel? In the future, I predict that robots will be used more and more for such tasks, saving their human operators many skinned knees and ruined T-shirts.

Of course, remote-controlled robots have limitations. A video camera can never give you a wide view like being there yourself, and the subtleties of the environment — especially odors — can't be sensed remotely. Also, the range of remote control is limited unless you have a high-powered, high-cost system such as the ones the government uses for unmanned vehicles. Still, these limitations will diminish as technology improves, thereby enabling remote-controlled robotics to serve humans in ever-more-helpful ways.

If you first feed all the RC channels into ARobot's controller, the program can decide whether to control the motors based on the RC receiver or perform intelligent, autonomous operation.

A good example of a situation where this would be desirable is a remote inspection application, in which the robot must navigate around rocks and other obstacles. If everything is going well, the robot's program simply receives commands from the RC receiver and controls the drive, steering, and camera motors accordingly. But if the robot were to run into a rock triggering one of the whiskers, the program could take over and stop the drive motor, reverse course, and steer away from the obstacle. When things look clear, control reverts to the operator.

Getting things wired

To get things set up to switch between these two operation modes, two more RC signals must be wired into the expansion board using port pins currently assigned to other devices, if you've completed earlier projects in this book. You might have to sacrifice the light sensor, PIR, or other device to gain the valuable extra port pins.

The steering and camera motor connectors should be disconnected from the remote-control system and connected to ARobot's main controller board, as they were before this project, to allow the program to control them.

Reading the other two signals is made simple for ARobot using the `pulsin` command, which is part of the PBasic language inside the Basic Stamp 2. You must set the pin number parameter to the number you choose to use for each motor. `pulsin` sets a variable of your choosing to a value somewhere between 500 and 1000, which represents the position of the control on the RC transmitter. This value tells the program what position the joystick control is in and can be used to control any device on ARobot.

The details of how the `pulsin` command works are described in the Basic Stamp 2 programming documentation. Trust me, it's not important that you completely understand how it works.

After the program receives the value from the `pulsin` command, you may want to *scale* it to make it easier to work with. For example, if you want to work with a value between 0 and 100 instead of 500 and 1000, use simple mathematical operations to get the number you want. In this example, you would need to first subtract 500 to get values from 0 to 500, and then divide by 5.

Software

Here's the code that gives you and your robot the best of both the RC and autonomous worlds. The code in Listing 17-2 shows you how to make your ARobot read three RC receiver channels on ports 6, 7, and 8, and scale them to values of 0 to 100.

Listing 17-2:   Reading Three RC Channels

```
rc1      var word            'RC Channel 1 value
rc2      var word            'RC Channel 2 value
rc3      var word            'RC Channel 3 value


         pulsin 6,1,rc1      'Read channel 1
         rc1=(rc1-500)/5     'Scale down value to 0-100

         pulsin 7,1,rc2      'Read channel 2
         rc2=(rc2-500)/5     'Scale down value to 0-100

         pulsin 8,1,rc3      'Read channel 3
         rc3=(rc3-500)/5     'Scale down value to 0-100
```

Now it's up to you to decide how to use the resulting variables to control the robot's functions. Hack away my fellow robot builder, and experience the inner joy of technological achievement! (In other words, go for it and have fun.)

Part V
The Part of Tens

The 5th Wave   By Rich Tennant

Well heck, Justin—that's darn impressive! What else can that little programmable robot of yours do? How about sewing up and dressing that incision?

In this part . . .

Everybody loves lists, and robot builders are no exception. The Part of Tens has two great little chapters that call out the best stuff in quick and easy-to-absorb lists.

In this part, you get to read about ten great suppliers of robot parts and ten important safety pointers to follow when building robots. Read these two chapters and you'll be well stocked but never shocked (or cut, or bruised, or whatever).

Chapter 18

Ten Excellent Parts Suppliers

To design and build a robot, you must have a good understanding of what parts are available and what they cost. Bottom line: Don't underestimate the value of a good supplier base to a successful robot-building experience.

In this chapter, I list my favorite parts suppliers. Those listed here should be able to provide most of the components you'll need for a robot project. In the last section, I include a list of smaller suppliers that carry some neat little items that just might come in handy.

Tower Hobbies

Products: RC servo motors, radios, hardware, and so on

Catalog: Yes

Online ordering: Yes

Web site: www.towerhobbies.com

Address: P.O. Box 9078, Champaign, IL 61826-9078

Phone: 217-398-3636

Fax: 217-356-6608

E-mail: info@towerhobbies.com

If you're looking for remote-control (RC) products, look no further than Tower Hobbies. They offer an incredible selection of products that will make any budding robot builder grin from gear to gear. Tower offers a good selection of RC servo motors and radios from Airtronics, Hobbico, Hitec, and Futaba, as well as miniature hardware from Du-Bro. Tower also carries a wide variety of kit planes, cars, and boats. They also offer other hobby-related products such as science kits.

Their Web site is easy to navigate and includes pictures for almost every product. Prices are clearly listed and an easy-to-use shopping-cart feature eases those multi-item purchasing binges.

Two other features that make Tower Hobbies stand out are their online tutorials dealing with RC systems and their exhaustive links section.

Supercircuits

Products: Video cameras, displays, RF, and accessories

Catalog: Yes

Online ordering: Yes

Web site: www.supercircuits.com

Address: One Supercircuits Plaza, Liberty Hill, TX 78642

Phone: 800-335-9777

E-mail: Online form

You just knew there'd be a store called *super* something in here, didn't you? Supercircuits is your one-stop source for all things video. They stock cameras, displays, transmitters/receivers, recorders, and accessories such as brackets and cables. They also offer a large range of video cameras including small board cameras that are perfect for robot projects.

The menus on the Supercircuits Web site make it easy to browse. The Web site also includes useful resources such as a lens calculator, a lux (a measurement of incident light) chart, and a camera comparison guide.

Mouser Electronics

Products: Electronic components, tools, and hardware

Catalog: Yes

Online ordering: Yes

Web site: www.mouser.com

Address: 1000 North Main Street, Mansfield, TX 76063-1514

Phone: 800-346-6873

Fax: 817-804-3899

E-mail: help@mouser.com

As the parent of multiple robots, I know I shouldn't have favorites, but I do. Mouser Electronics is by far my favorite supplier for everyday electronic items such as resistors, capacitors, transistors, and ICs as well as hardware such as small screws and washers. They also carry a decent assortment of tools such as soldering irons, voltmeters, and even a few sheet-metal machines. The great thing about Mouser is that they stock almost every type of product, and they're definitely hobbyist-friendly with their no-minimum-order policy.

Mouser has a great catalog, but their Web site's search capability will become your best friend. When you search for and then locate a part, possible substitutions and related products are also listed along with pricing for various quantities and the number of items currently in stock. If you're working with electronics, you probably won't need any catalog but Mouser's.

Parallax

Products: Basic Stamp Microcontrollers, kits, and so on

Catalog: Yes

Online ordering: Yes

Web site: www.parallax.com

Address: 1000 North Main Street, Mansfield, TX 76063-1514

Phone: 916-624-8333

Fax: 916-624-8003

E-mail: eminfo@parallax.com

Parallax is the producer of the hugely popular Basic Stamp microcontrollers (such as the one used in ARobot). The Basic Stamp product line now includes several different models that offer different speeds, memory capacities, programming languages, I/O ports, and other features.

These little microcontrollers have single-handedly changed the landscape for robot builders over the last decade by providing an easy-to-use, self-contained controller. The controllers are easy to program and are capable of running many small applications.

The Parallax Web site offers an abundance of product information, including a comparison chart and data sheets for each product. A wealth of application and tutorial information is also provided.

McMaster-Carr

Products: Hardware, tools, and materials

Catalog: Yes, but it's hard to get

Online ordering: Yes

Web site: www.mcmastercarr.com

Address: P.O. Box 4355, Chicago, IL 60680-4355

Phone: 630-833-0300

Fax: 630-834-9427

E-mail: chi.sales@mcmaster.com

If you visit a robot-building buddy's workshop and spot a yellow catalog that's bigger than the New York City phone book (volumes 1 and 2), it's probably the McMaster-Carr catalog. These folks have everything in the way of hardware, tools, and materials. Any robot builder will be drawn to the large sections on fasteners, aluminum stock, and hard-to-find products such as rubber bumpers and specialty hinges.

Through the wonder of technology, McMaster-Carr has managed to put their giant catalog, complete with pictures and product dimensions, online. You'll get a taste for the number of offerings when you visit their home page, which includes a seemingly endless index of products.

McMaster-Carr isn't using smoke-and–mirrors, however: They stock everything in the catalog and it's not uncommon for the products I order to appear at my door before I get around to removing the fax from the paper tray.

Stock Drive Products

Products: Bearings, gears, pulleys, chains, collars, and so on

Catalog: Several

Online ordering: Yes

Web site: www.sdp-si.com

Address: 2101 Jericho Turnpike, Box 5416, New Hyde Park, NY 11042-5416

Phone: 516-328-3300

Fax: 516-326-8827

E-mail: support@sdp-si.com

There are plenty of suppliers of gears, pulleys, bearings, and those other small precision parts needed to add some *go* to your robotic creation, but no one offers the range of products that Stock Drive does. Luckily, they've broken their product line down into separate catalogs to make your journey easier — and the catalogs contain a wealth of reference material as well.

Unlike many other miniature drive component suppliers, Stock Drive Products has a great Web site that provides online ordering, selection matrices, and downloadable data sheets. All things considered, SDP is such an excellent source for small mechanical components that, if you can't find it at SDP, you should probably change your robot's design.

You may find that some of the functionality on the Stock Web site is missing on non-IE browsers. Also, their marketing department has chosen to use annoying pop-ups.

Radio Shack

Products: Basic Electronic components, tools, and so on

Catalog: Yes

Online ordering: Yes

Web site: www.radioshack.com

Address: 100 Throckmorton Street, Fort Worth, TX 76102

Phone: 817-415-3011

E-mail: Online form

Radio Shack is on practically every corner, with thousands of stores in the United States and many overseas. Most people know them for batteries and walkie-talkies, but robot builders know them as a source for that resistor or connector needed at the last-minute to finish a product.

Although Radio Shack's line is not as comprehensive as Mouser, they do carry a basic range of components including capacitors, resistors, transistors, ICs, connectors, wire, and fuses. They even carry a decent range of low-cost tools such as soldering irons.

Radio Shack has an excellent Web site with online ordering and a good search feature. Also provided is a handy store locator, coupons, and manuals for various products.

80/20

Products: Aluminum framing components, brackets, and so on

Catalog: Yes

Online ordering: Yes

Web site: www.8020.net

Address: 1701 South 400 East, Columbia City, IN 46725

Phone: 260-248-8030

Fax: 260-248-8029

E-mail: info@8020.net

Here's a company that not only has an interesting name, but also a great product line. 80/20 offers mechanical building components for those bigger robot projects including aluminum bars, brackets, and accessories such as special hinges and sliding rails. All components bolt together with unique hardware that makes building fun. They also offer cutting and drilling services for the tool challenged.

80/20 products are great for building industrial automation products but should also be useful for constructing large robots. Request their catalog and visit their Web site for inspiration and part numbers.

Edmund Scientific

Products: Robot kits, science kits, and components

Catalog: Yes

Online ordering: Yes

Web site: www.scientificsonline.com

Address: 60 Pearce Ave., Tonawanda, NY 14150-6711

Phone: 800-728-6999

Fax: 800-828-3299

E-mail: scientifics@edsci.com

Started in 1942 with a line of optical components, Edmund Scientific has become one of the best-known sources for science kits, lenses, test equipment, and components used in laboratories and education.

Edmund Scientific — which is now a division of Scientifics, the world's largest science kit producer — offers a nice range of robot kits and components for the budding robot builder. Their Web site provides thorough descriptions of each product and a nicely organized shopping-cart feature for online ordering.

Parts Suppliers a la Carte

This section includes a potpourri of more specialized parts sources. Some of these companies are not really small; they simply offer unique items that aren't easy to find. Others are small and specialized.

Most of the companies listed here offer a wide variety of electronic components, tools, computer parts, sensors, motors, and mechanical devices. All offer a catalog, so I encourage you to contact them to build up your robot-building bookshelf.

Circuit Specialties

Products: Electronic components, computer parts, and so on

Web site: www.web-tronics.com

Phone: 800-528-1417

Jameco Electronics

Products: Electronic components, computer parts, and so on

Web site: www.jameco.com

Phone: 800-831-4242

Marlin P. Jones & Assoc.

Products: Electronic components, computer parts, and so on

Web site: www.mpja.com

Phone: 800-652-6733

All Electronics

Products: Electronic components, computer parts, and so on

Web site: www.allelectronics.com

Phone: 888-826-5432

Mendelsons Electronics

Products: Electronic components, computer parts, and so on

Web site: www.meci.com

Phone: 1-800-344-4465

Ramsey Kits

Products: Electronic kits, tools, components, and so on

Web site: www.ramseyelectronics.com

Phone: 800-446-2295

Small Parts

Products: Mechanical components, material, tools, and so on

Web site: www.smallparts.com

Phone: 800-220-4242

JK Micro

Products: Single board computers

Web site: www.jkmicro.com

Phone: 530-297-6073

Carl's Electronics

Products: Electronic kits and robot kits

Web site: www.electronickits.com

Phone: 800-439-1417

Chapter 19

Ten Safety Pointers

In This Chapter

▶ Observe reasonable precautions
▶ Think ahead
▶ Use the right tools

Compared to defusing bombs and walking tightropes, building robots is a safe endeavor. Still, I've listed some safety tips here that may come in handy.

In addition to the specific pointers included here, remember that working with tools is a major source of accidents, so be sure to use eye protection and follow the manufacturer's usage instructions. But in general, if you use common sense and the following advice, you can avoid almost all accidents.

Cut Away from Your Body

The advice to cut *away* from your body may seem obvious, but for some reason it's not natural because cutting towards yourself tends to give you better leverage. Nevertheless, you should always point that sharp cutting thing away from your body. Having a knife slip while cutting is a common event, so make sure when you do slip that the blade goes away from anything delicate, especially your skin! Apply this rule to razor-sharp hobby knives and dull pocketknives alike.

Avoid the Pinch Points

This next injury is a common one that almost everyone who works with robots has experienced, yet I've never seen it mentioned in the books I've read. The main culprit here is needle-nose pliers. When these pliers are open, there's usually a space at the joint. When you close the pliers, the two sides of the tool come together. This point is very near the handles, and fingers tend to slip up into the open space and then get caught when the pliers close. The result is a pinched finger and some language my publisher will not let me repeat here. So, keep those digits clear because you'll need them for future robotics projects.

Slipping Is Bad

Sometimes when you're applying a lot of force to a tool it slips away from you. Because the nature of slipping suggests a certain loss of control, where the tool lands and what damage it may do is unpredictable, to say the least.

I've found that the main slipping culprit is the lowly screwdriver. Make sure to use the proper size and type of screwdriver for the job, and be sure you're in the correct position to apply the force needed to complete the job.

A close cousin to slipping is dropping. Having a hot soldering iron, an electric drill, or the infamous hobby knife fall off your workbench and onto your foot is bad form and could possibly damage the valuable tool beyond repair — oh, and it might hurt too!

Soldering Pitfalls

Soldering is pretty easy and fairly safe, but it does involve high temperatures, so just about everyone gets burnt once or twice.

Soldering irons usually have cords that attach to the base station and these cords can get tangled in your legs and in the arms of chairs. The result is a hot falling tool that can burn the carpet or worse.

Solder itself can drip off the tip of the soldering tool, so make sure you have a suitable work surface to catch the drips with minimal damage, and don't wear shorts unless you're fond of artificial freckles on your thighs.

Hot Glue Follies

Hot glue is a fabulous invention, right up there with duct tape, chewing gum, and coat hangers. However, hot glue has this tendency to drip and hit any soft skin-like surface. When it does, it tends to stick faster than you can say "ouch!" The result is a minor burn that commonly leaves a mark in the shape of a midwestern state on your skin. To prevent this, take care with the gun, and make sure you have a stable, non-flammable surface to lay the gun down on when it's not in use.

Super Glue on You

Super glue is very strong and can be more dangerous than a power tool. This type of glue is incredible at attaching skin together — in fact, that's what super glue was made for, but it's not what it's sold for. Don't let the glue touch your skin. If it ignores your caution and does anyway, don't touch anything else or you'll just involve some other poor creature or plastic part in your dilemma.

The solution? When you purchase the glue, also purchase a bottle of debonder, which is the best way to handle the problem.

Dancing around the Drill

You don't need to hear horror stories from machine shops about the problems a power tool can cause, but let me give you a few words of advice about the common activity of drilling.

First, don't wear long sleeves while drilling because they can get caught in the drill bit and make a mess of your arm. Remove any loose clothing and any nearby cloth or material.

Second, clamp what you're drilling tightly. Don't rely on your hand to keep that piece of sheet metal from grabbing the drill bit and starting to spin.

Third, don't use power tools alone if you have a buddy handy. It's safer to have someone there to help and to watch.

Fourth, and certainly not last, wear safety goggles. They aren't nearly as un-cool as an eye patch.

AC Stands for <u>A</u>re You <u>C</u>razy!

Many an accident is caused when you underestimate a powerful force, only to learn later what damage it can cause. The voltage on an AC line is powerful and can hurt you, or at least scare the heck out of you. Don't ever work on a device that's plugged in. (Luckily, most mobile robots work on much lower voltages, so you're usually pretty safe.)

Discharging Capacitors

After a device is turned off, the capacitors in the circuit can retain a charge. This is especially true of televisions and other high-power devices. It's best to let the device sit for a while or to manually drain the capacitors with a power resistor.

Clipping Nippers

Wire cutters (sometimes called nippers, dikes, or diagonal cutters) are useful, and everyone should have a pair or two, but some potential problems are associated with their use. First, their blades are usually metal, which means that whatever wire you're cutting will become electrically connected to the handles, and although the handles are insulated, they may connect your skin to the circuit.

Second, if you're cutting two wires at a time, those two wires become connected during the cutting process, and that could cause a shorted battery or worse.

Finally, nippers are sharp and can trim more than a fingernail, so follow all my earlier advice in this chapter about playing with sharp objects.

Index

• M •

FOR DUMMIES®

A world of resources to help you grow

TRAVEL

0-7645-5453-0

0-7645-5438-7

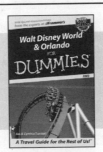

0-7645-5444-1

Also available:

America's National Parks For Dummies
(0-7645-6204-5)

Caribbean For Dummies
(0-7645-5445-X)

Cruise Vacations For Dummies 2003
(0-7645-5459-X)

Europe For Dummies
(0-7645-5456-5)

Ireland For Dummies
(0-7645-6199-5)

France For Dummies
(0-7645-6292-4)

Las Vegas For Dummies
(0-7645-5448-4)

London For Dummies
(0-7645-5416-6)

Mexico's Beach Resorts For Dummies
(0-7645-6262-2)

Paris For Dummies
(0-7645-5494-8)

RV Vacations For Dummies
(0-7645-5443-3)

EDUCATION & TEST PREPARATION

0-7645-5194-9

0-7645-5325-9

0-7645-5249-X

Also available:

The ACT For Dummies
(0-7645-5210-4)

Chemistry For Dummies
(0-7645-5430-1)

English Grammar For Dummies
(0-7645-5322-4)

French For Dummies
(0-7645-5193-0)

GMAT For Dummies
(0-7645-5251-1)

Inglés Para Dummies
(0-7645-5427-1)

Italian For Dummies
(0-7645-5196-5)

Research Papers For Dummies
(0-7645-5426-3)

SAT I For Dummies
(0-7645-5472-7)

U.S. History For Dummies
(0-7645-5249-X)

World History For Dummies
(0-7645-5242-2)

HEALTH, SELF-HELP & SPIRITUALITY

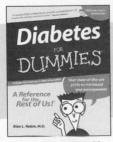

0-7645-5154-X

0-7645-5302-X

0-7645-5418-2

Also available:

The Bible For Dummies
(0-7645-5296-1)

Controlling Cholesterol For Dummies
(0-7645-5440-9)

Dating For Dummies
(0-7645-5072-1)

Dieting For Dummies
(0-7645-5126-4)

High Blood Pressure For Dummies
(0-7645-5424-7)

Judaism For Dummies
(0-7645-5299-6)

Menopause For Dummies
(0-7645-5458-1)

Nutrition For Dummies
(0-7645-5180-9)

Potty Training For Dummies
(0-7645-5417-4)

Pregnancy For Dummies
(0-7645-5074-8)

Rekindling Romance For Dummies
(0-7645-5303-8)

Religion For Dummies
(0-7645-5264-3)

Available wherever books are sold. Go to www.dummies.com or call 1-877-762-2974 to order direct

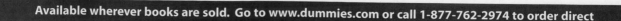

Special Offer. Save $20.00!

ARobot Kit Rebate

For information about purchasing the ARobot kit, including pricing and availability, visit our Web site at www.robotics.com or call us at 817-571-4528. (Price and availability subject to change without notice.) The Basic Stamp 2 controller used with ARobot is sold separately.

To receive your $20.00 rebate on the purchase price of ARobot, complete this form and mail it, along with the proof of purchase, to the following address:

Arrick Robotics
P.O. Box 1574
Hurst, TX 76053

Name: _____

Address: _____

City: _____ State: _____ Zip: _____

Special Offer. Save $6.00!

Soccer Jr. Kit Rebate

Save $6.00 when you purchase the Soccer Jr. robot kit from
www.HobbyTron.com.

To redeem your $6.00 coupon on the purchase price of Soccer Jr., use the following coupon code when you are in the shopping cart section of our Web site:

dummyrobot

For additional information about purchasing Soccer Jr., including pricing and availability, visit our Web site at www.HobbyTron.com/soccerjr.html. (Price and availability are subject to change without notice.)